# Prophet of the Modern Technological Age

**By Michael W. Simmons**

Copyright 2016 by Michael W. Simmons

Published by Make Profits Easy LLC

Profitsdaily123@aol.com

facebook.com/MakeProfitsEasy

# Table of Contents

Introduction ............................................................. 4

Chapter One: The Inventor's Early Life ............ 25

Chapter Two: The Struggle for Recognition ...... 58

Chapter Three: AC versus DC ............................. 81

Chapter Four: Dreams and Visions ................... 101

Chapter Five: The Wizard of Fifth Avenue ...... 124

Chapter Six: Houston Street and Beyond ........ 148

Chapter Seven: Colorado ................................... 166

Chapter Eight: Radio .......................................... 179

Chapter Nine: Decline and Fortune ................. 199

Works Referenced ............................................... 239

# Introduction

"My belief is firm in a law of compensation. The true rewards are ever in proportion to the labor and sacrifices made. This is one of the reasons why I feel certain that of all my inventions, the Magnifying Transmitter will prove most important and valuable to future generations. I am prompted to this prediction not so much by thoughts of the commercial and industrial revolution which it will surely bring about, but of the humanitarian consequences of the many achievements it makes possible. Considerations of mere utility weigh little in the balance against the higher benefits of civilization. We are confronted with portentous problems which can not be solved just by providing for our material existence, however abundantly. On the contrary, progress in this direction is fraught with hazards and perils not less menacing than those born from want and suffering. If we were to release the energy of atoms or discover some other way of developing cheap and unlimited power at any point of the globe this accomplishment, instead of being a blessing, might bring disaster to mankind in giving rise to dissension and anarchy which would ultimately result in the enthronement of the hated regime of force. The greatest goodwill comes from technical improvements tending to unification and harmony, and my wireless transmitter is preeminently such. By its means the human voice and likeness will be reproduced everywhere and factories driven thousands of miles from waterfalls furnishing the power; aerial machines will be propelled around the earth without a stop and the sun's energy controlled to create lakes and rivers for motive purposes and transformation of arid deserts into fertile land. Its introduction for telegraphic, telephonic and similar uses will automatically cut out the statics and all other interferences which at present impose narrow limits to the application of the wireless."

Nikola Tesla, *My Inventions*

The really extraordinary thing about Nikola Tesla is that his experiments in electricity hold the power to fascinate even those of us who have little background in science and no real understanding of the principles of electrical engineering. Reading descriptions of Tesla's electrical demonstrations and looking at photographs of his generated lightning can help the modern reader, who grew up in a world where electricity is so ordinary as to be unremarkable and invisible, to begin to appreciate what it was like when humans had only just begun learning how to harness electricity's power.

In the early 1890's, when Tesla stunned the world with a series of lectures and demonstrations showcasing his electrical experiments, electricity was a phenomenon that was little understood even by the first minds in the scientific field. Electricity still played only a negligible role in the life of the average person. If you lived in New York, Chicago, London, Paris, or another large city in America or western Europe, it is possible that you would have encountered electric lighting in the streets, or electrically powered trolley cars—but only the very wealthy had electric lights in their own homes.

For the common person, there was an aura of mystery and fascination around electricity that was only heightened by the rumors about how dangerous it was. Early direct current electrical power *could* be quite dangerous—rudimentary wiring systems were prone to short circuits, explosions, and fires. Sometimes horses would bolt in the streets, because they received electrical shocks through their metal shoes when walking down a section of pavement close to a transformer. But the aura of danger was also a result of false propaganda: Thomas Edison, America's most famous electrical engineer, was the bitter enemy of Tesla's alternating current system (and, over time, of Tesla himself). Despite the fact Tesla had engineered his electrical induction motor specifically to provide a safer alternative to direct current electrical power, Edison was committed to convincing the American public that alternating current electricity was deadly. During the so called "War of the Currents", a war waged almost single-handedly by Edison, he and his assistants held public demonstrations in which they attempted to prove the deadliness of alternating current electricity by electrocuting animals in public— sometimes livestock, other time cats and dogs that had been stolen from people's yards by schoolboys Edison hired. All of this, just to hammer home his point that electricity could kill—unless it was *his* kind of electricity. Edison went so far in this campaign that it was as a direct result of his lobbying when the first death row inmate in an American prison to be executed by electrocution was killed by an alternating current.

Edison's propaganda had two purposes: to insure the superiority of his own direct current system, and to crush the competition posed by Tesla's alternating current system. And he succeeded beautifully for a few years; belief in the dangerousness of electricity did not stop the White House from installing electric lighting, but it did mean that the President was not personally allowed to turn the light switches, in case of a short circuit. But alternating current was the superior system, and in time, even Edison was forced to acknowledge it, though not until long after it had begun to dominate the market. He never acknowledged how he had falsified reports of fatal accidents from alternating current electricity, however, or any of the rest of his false propaganda.

The chief difference between Edison and Tesla was that Edison was not just an electrical engineer. He was an American entrepreneur in the Industrial Age, an era in which millionaire robber barons were creating vast monopolies by buying up and consolidating coal, oil, steel, any commodity that stood a chance of making them a lot of money over a period of many years. Ethical considerations held no special sway over industrialists like Vanderbilt, Astor, Carnegie, Morgan, Van Allen—they made liberal use of threats, intimidation, bribes, and even violence to weed out competitors and consolidate their power over the markets. Edison was trying to advance in the same marketplace. A self-made man with no formal education, Edison was a practical inventor with no use for theory. He saw his electrical inventions as a route to fame, fortune, and legacy, and it didn't especially matter to him whether his system or Nikola Tesla's system was better—he just wanted his system to win. In both the best and worst of ways, Edison represented the nineteenth century ideal of the self-made man who had pulled himself up by his boot straps to achieve the American dream.

Nikola Tesla represented something quite different. Born in Croatia, educated in German language schools and Austrian polytechnic academies, Tesla was nonetheless largely self-taught when it came to his life's work of electrical engineering. But he was steeped in a culture that was far removed from the American race to improve one's fortunes and elevate one's station in life at any cost. Tesla was a highly cultured person. From his parents, he learned to recite lengthy epic poems from memory. He appreciated music, literature, fashion, and fine dining, not merely as the trappings of financial success and social status, but in themselves. He contributed more to the field of electrical engineering than any single individual before or after him, but he was much more than just an electrical engineer. Perhaps because of his deep connection to the arts and humanities, or because of his essentially poetic temperament, Tesla never saw his scientific discoveries as merely marketable products that he could sell to industrialists for large sums of money and profitable patent royalties. Tesla's relationship with electricity was partly that of an artist to his medium, partly that of priest to the eternal spiritual truths he wished to impart to his flock. He saw in his experiments, his machines, and his devices a means to make the world better, to end the problems and sufferings of humanity. And on an even more basic level, Tesla delighted in using electricity to awe and delight all who witnessed his demonstrations.

Handsome, tall, and slender, well groomed and well dressed, articulate in a number of languages, Nikola Tesla cut a striking figure during his many lectures and demonstrations. Dressed in white tie and tails, Tesla resembled our image of what a magician should look like far more than our idea of what a scientist should look like. He was a showman as much as an instructor or a lecturer. And the demonstrations of electrical power he gave to admiring audiences seemed indeed to mimic the magical powers of a wizard who commanded lightning from the heavens. Even his explanations of the experiments he performed on stage sounded more like the patter of a sideshow magician than the dry, academic, technical explanations we tend to associate with scientists expounding on their craft. The following excerpt from his lecture to the Institute of Electrical Engineers in London demonstrates his ability to mesmerize with words as well as actions:

"Here is a coil which is operated by currents vibrating with extreme rapidity, obtained by disruptively discharging a Leyden jar. It would not surprise a student were the lecturer to say that the secondary of this coil consists of a small length of comparatively stout wire; it would not surprise him were the lecturer to state that, in spite of this, the coil is capable of giving any potential which the best insulation of the turns is able to withstand: but although he may be prepared, and even be indifferent as to the anticipated result, yet the aspect of the discharge of the coil will surprise and interest him. Every one is familiar with the discharge of an ordinary coil; it need not be reproduced here. But, by way of contrast, here is a form of discharge of a coil, the primary current of which is vibrating several hundred thousand times per second. The discharge of an ordinary coil appears as a simple line or band of light. The discharge of this coil appears in the form of powerful brushes and luminous streams issuing from all points of the two straight wires attached to the terminals of the secondary."


Tesla gave these magical-seeming demonstrations before lecture audiences in London, Paris, New York, Chicago, Belgrade, and elsewhere around the world. As a lionized member of New York high society in the 1890's, he also gave demonstrations to private audiences, composed mostly of celebrities and wealthy potential patrons, in his private laboratories in Manhattan following dinner parties, as a sort of after dinner entertainment. Indeed, Tesla was so much a part of the nation's most rarefied social circles that it engendered resentment and spite amongst other, less brilliant and less famous inventors, who often found a sympathetic audience for their anti-Tesla newspaper articles ready made for them by Edison's relentless campaigning in the early 1890's. But this was perhaps an inevitable consequence of how famous and well-loved Tesla was in general. He was a celebrity on a level that no scientist could probably imagine today; our best known scientists, like Stephen Hawking and Neil DeGrasse Tyson, are known for being educators first and foremost. Tesla was something much more modern: he was a spectacle. And nowhere, perhaps, was Tesla's scientific magic displayed to better advantage than at the Chicago World's Fair in 1893.

The Chicago World's Fair of 1893, also called the Columbian Exposition in honor of Christopher Columbus, was designed to be a grand spectacle such as the world had never before seen. It took place in the midst of a financial recession, in a city where the poorest citizens stood in bread lines, but it was meant to reflect the grandeur of America's industrial pre-eminence in the world, and to showcase how the nation had transformed itself from the wild, barely civilized frontier nation of the decades after the Civil War into one of the world's leading financial powers, a land of opportunity where immigrants flocked from around the globe in search of a new, better life.

If Edison represented one version of the American dream, the poor boy who had pulled himself up by his bootstraps to capitalize on his

God given talents through hard work, Tesla represented a different, but equally authentic version: the brilliant immigrant who traveled across an ocean, full of brilliant ideas that made him a poor fit for the culture of the Old World, to plug his imported genius into the switchboard of the nation's vibrant growing technological industries. At the World's Fair of 1893, this was true on a literal level: the fair displays were designed to showcase the power of electricity, audiences glimpsing for the first time in history what is to us today that most common of spectacles, a city shining in the darkness because it has been lit with electrical lights. And all of that electricity was running on Tesla's first and most cherished invention, the alternating current induction motor with rotating magnetic fields that Thomas Edison had first rejected, then attempted to malign and destroy.

Tesla's inventions lit the Columbian Exposition, but that was not the only role he played there. In the enormous Electricity Building, Tesla had his own display rooms, where he showcased his myriad inventions, the likes of which had never before been seen by the ordinary men and women who came to gawk at them. Tesla's chief joy in life was shutting himself away in his private laboratories to pursue theories and perform experiments strictly for the joy of seeing what he could do—much to the dismay of his investors, he vastly preferred this to working on inventions with obvious industrial applications or marketing potential. Like an artist, he was content to shape electricity in ways that were merely fantastic, fascinating, and beautiful, and the World's Fair exhibit gave him a chance to astonish the uninitiated with these displays. He demonstrated the first phosphorescent tube lights, delicate hand blown tubes in the shapes of letters that spelled out phrases—the nineteenth century precursor to the fluorescent and neon lights that advertise business and slogans today. And this was only the most mundane of his displays. Using electricity, Tesla made objects spin in the air, made sparks jump from his hair and clothing, surrounded himself in a cold fire that did not burn, and made lightning leap from place to place. The people of Chicago filing through his display rooms were uncertain of what, precisely, they had just seen, but they knew that it was unlike anything they had ever seen before.

**Tesla the Visionary**

One of the most remarkable things about Tesla is how far into the future his vision reached. Not only did he invent things for which the application would not become obvious until other fields of science had caught up with him, decades later, but he intuited the possibility and inevitability of many technologies that would not come into existence until the late twentieth or early twenty first centuries. It raises the fascinating question of what inventions inventors may devise in the future that will have been anticipated by Tesla as well. More than one of his inventions which were dismantled and lost to history have been described as having effects that scientists cannot presently account for or reproduce. Margaret Cheney writes of some of the technologies he envisioned—some of which will sound familiar to the contemporary reader:

"He began to achieve effects with high-voltage equipment that opened an infinity of possibilities. By learning to create artificial lightning he hoped not only to discover how to control the world's weather but also how to transmit energy without wires. And this in turn meshed with research that he hoped would enable him to build the first world-wide broadcasting system."

Scientists in the early twenty first century have yet to devise a means of controlling the weather, or lighting the entire world by the illumination of atmospheric gases—Tesla's "terrestrial night light"—though who knows what the future may hold? But his inventions continue to affect our every day lives in ways other than the obvious. Just to take one example at random, if you have ever smiled at seeing one of those little plastic flowers in a pot that "dance" back and forth when placed in the sunlight, then Tesla's inventions have added to the small joys of your life; it is his design for heat-powered evacuated bulbs that powers these toys, and he designed them for not other purpose than to please a friend of his named Katharine Johnson.

Tesla's inventions exist on every level of our technological infrastructures. Unlike Thomas Edison, it is unusual to hear his name spoken in an elementary school history lesson, but his influence was pervasive, extending into all areas of our modern lives. And the technology of the future is increasingly inspired by his example and increasingly dependent on his prototypes. In 2003, the Tesla Motor Company was formed in order to develop an affordable electric powered car that runs off the very alternating current induction motor that Tesla first patented in 1888.

Though celebrated and admired in his lifetime, Tesla's reputation faded into obscurity for many decades after his death, as Thomas Edison's story became the dominant narrative of scientific experimentation and business success in the American fable. It seemed not to matter that Tesla had looked into the future, that he had dreamed ahead of his time; he was forgotten for the simple fact that he had not formed a thriving business corporation with his name on it that would have a vested financial interest in defending his legacy into the far future. Edison's company, General Electric, is still a household name, and bizarrely, their marketing departments continue to make passive aggressive jabs at Tesla's memory, as if the War of the Currents was still raging.

It wasn't that Tesla's inventions didn't make money; on the contrary, they made millions of dollars—hundreds of millions in today's currency. But Tesla was not a businessman. His mind wasn't on the task of zealously protecting his financial interests. He let others form the corporations—some with his name on them, some without. He attempted to protect his patents, because it did matter to him very much if someone else attempted to take credit for what he had built, but he did not protect his royalties to the same degree. In fact, he simply forfeited his claims to over twelve million in royalties from his induction motor, because he believed that doing so was necessary to saving a friend's company. These are the kinds of weaknesses—caring less for business success than for using his inventions philanthropically to improve life on earth—for which American history, until recently, tends to consign people to oblivion after their death.

But Tesla's reputation has been enjoying a considerable resurgence in recent decades. New documentaries have been made, new biographies have been written, and new companies have been formed to honor him—America has rediscovered its lost wizard. And in this book, we will be exploring a few of his most famous magical—scientific—tricks. As Arthur C. Clarke wrote, any sufficiently advanced technology is indistinguishable from magic; and Tesla was more advanced than virtually anyone before or since.

# Chapter One: The Inventor's Early Life

**The Tesla Family**

In the middle of the nineteenth century, shortly before the outbreak of the Civil War in faraway America, and the close of the Crimean War between Russia and the Ottoman Empire, Nikola Tesla was born in a small village called Smiljan, in Croatia, to a Serbian family. He was born precisely at midnight in between June 9 and June 10, 1856, as the fourth child of Milutin Tesla, a priest of the Serbian Orthodox Church, and his wife, Duka Mandíc. His eldest sibling was a brother, named Dane, sometimes written as Daniel. He had two older sisters, Milka and Angelina, and younger sister, Marica.

Tesla left behind a number of volumes of his own writing, and on the whole, they are more scientific volumes than biographical or literary documents, as they were intended to explain the reasoning behind the many devices he imagined and built. But in his autobiography, titled *My Inventions*, he shares such details about his childhood, education, and early life as he considers relevant to the understanding of his scientific accomplishments. In his book, Tesla describes a childhood rich with intellectual stimulation. Tesla's family was of a social class that traditionally educated their children for careers in either the army or the church. Milutin Tesla had been intended for an army officer, but had left the officer's training academy to become a priest in the Serbian Orthodox Church, and he wished for his son Nikola to become a priest as well.

To prepare him intellectually for this profession, Tesla's father gave him problems to solve, and enormous memorization tasks to complete. Throughout his life, one of Tesla's most significant intellectual attributes was his prodigious memory. He was able to memorize a sheet of paper covered in writing, or even in complicated mathematic and scientific formulas, after scarcely more than glancing at it. When designing his inventions, Tesla rarely drew up plans on drafting paper, as scientific engineers generally do, because he was able to visualize his designs in his mind in such profound detail that he could detect flaws in his mental blueprints without having to commit them to paper. It was not until late middle age that his memory became slightly less retentive, requiring him to occasionally work problems out with pen and paper.

Attempting to account for his extraordinary feats of memorization, Tesla credited his mother for passing the necessary raw genetic materials for his achievements down to him. Like many women of her ethnicity and social class, Duka Mandíc had never been taught to read or write—

when she was a child, her own mother had gone blind, and Duka, as the eldest daughter, had been compelled to take over her domestic duties, managing the household and caring for her many younger siblings. However, she had an avid appetite for knowledge and literature, and was capable of committing huge stores of information to memory. She could recite entire volumes of poetry by heart. Furthermore, Tesla saw in his mother the capacity for invention that defined his own career. Describing how his mother produced clothing for the family by planting seeds, growing plants, separating fibers, weaving fabric, and cutting and sewing clothing from the fabric, Tesla mourned that his mother had not lived at a time and place where women were given proper recognition for their abilities. In his mother's household management routine, Tesla saw her create many small inventions that made her home run smoothly, inventions which, with a proper scientific education behind them, might have translated to great accomplishments.

## Dane Tesla

While Tesla seems disinclined to characterize his childhood as an unhappy one, he suffered a severe trauma at an early age when his oldest sibling and only brother, named Dane, died in a violent accident at the age of twelve, when Nikola was five. The precise circumstances of Dane Tesla's death are somewhat mysterious. In his autobiography, Tesla attributes the accident to a beautiful Arabian horse which had been given as a gift to his father; the horse was a beloved family pet, which Tesla describes as possessing an "almost human intelligence", which had saved his father from death during a ride through the forest in which Tesla's father was attacked by wolves. Tesla does not provide details as to how the accident that resulted in his brother's death occurred—whether the young Dane attempted to ride the horse and was thrown, or trampled, or suffered some other accident. In some accounts, the five year old Nikola was responsible for spooking the horse while Dane was riding it. Later in life, Tesla claimed that Dane died of his injuries after a fall down the stairs of the family's cellar. The one consistent element of the story of his brother's death is that the child Tesla witnessed the accident.

Whatever the precise circumstances surrounding the fatal accident may have been, his older brother's death had a profound effect on Tesla as he grew up. He describes his brother as "gifted to an extraordinary degree - one of those rare phenomena of mentality which biological investigation has failed to explain." As Tesla grew older and started his formal education, his own extraordinary abilities rapidly became evident. He became fluent in a number of languages at a young age, including English; but his principle and outstanding talent was in mathematics. His abilities in mathematics were so pronounced that his teachers nearly failed him in that subject because they were convinced he was cheating. This was because his ability to visualize the workings of a problem were so great that he never showed his work, or required his teachers to explain a problem after they had finished copying it out for him.

According to Tesla, however, his talents did not translate into happiness at home. By his own account, his extraordinary abilities served only to remind his parents of the brilliance of the son who had died, and to increase their sorrow over his death. Tesla wrote that he had little confidence in his talents growing up, because "anything I did that was creditable merely caused my parents to feel their loss more keenly." This, of course, is only the version of events that the adult Tesla remembered many years after they had happened. Whether or not his parents genuinely thought little of Tesla's abilities in comparison to those of the son who had died before he could put his own talents to the test as an adult, only the elder Teslas knew for sure.

However, even though Nikola Tesla undoubtedly grew up judging himself according to an impossible standard created by a dead older brother, it seems not to have given him bitter memories of his family life. He describes his father as a fairly gentle person who only resorted to corporal punishment once in all of Tesla's childhood, which showed a remarkable kind of parental forbearance by the standards of nineteenth century child-rearing norms. The one time that Milutin Tesla was driven to hit his young son, it was because Nikola had jumped onto the long, glittering dress train of one of Milutin's wealthiest parishioners, causing it to be torn away from the rest of the garment, and even then, Tesla described the blow as gentle, intended to shame him rather than to inflict physical damage.

## The Question of Education

The greatest conflict between Tesla and his father arose over the question of Tesla's future career. Tesla wished from an early age to be a trained as an engineer, a career aptly suited to his profound mathematical capabilities. Tesla's father was himself something of an inventor around the house—just as Duka Mandíc employed small inventions to enliven her domestic routines, Milutin Tesla was capable of tinkering with mechanical devices and tools to improve their performance. Tesla's own aptitude for engineering was demonstrated when, as a boy, his small village pooled the necessary funds to acquire a state of the art fire engine, complete with uniforms for the volunteer fire brigade members to wear. When the village of Smiljan assembled to see the fire engine in action, however, no water came out of the hose, to everyone's great disappointment. Tesla, however, correctly deduced that the engine's hose had collapsed in the river, and ran straight into the water to remove the obstacle. Water accordingly gushed from the engine's hose right into the faces of the concerned villagers who were examining it. According to Tesla, everyone was so delighted that he became the hero of the village for the day, and was carried around on the shoulders of the crowd.

Despite his obvious talent for engineering, however, Tesla's father was determined that Nikola would be a clergyman, like himself, and for many years he would not listen to his son's protests that he would undoubtedly be miserable as a Serbian Orthodox priest. It was not until 1873, when Tesla was seventeen, that his father changed his mind. Tesla had just completed his pre-university education at the gymnasium (a name for a school in eastern Europe and Russia in the 1800's). It was a period of study that normally took four years, although Tesla had completed it in three. Just before he was due to begin his seminary training, Tesla contracted a severe case of cholera during a return visit to his home village, and it came very close to killing him. As Tesla lay in bed, close to death, he mentioned to his father, perhaps jokingly, that maybe he would live if his father would allow him to study engineering like he'd always wanted. The doctors had all but given up on him at this point. His father replied that if he survived, he would send Nikola to the best polytechnic school in the world. Almost miraculously, Tesla began to recover immediately.

**"Luminous Phenomenon"**

In 1874, Tesla was notified of his conscription into the Austro-Hungarian army for a term of service that would have lasted three years. The prospect of military service was the only thing more repugnant to Tesla than a career in the priesthood. The precise details of the arrangement are unknown, but it appears that his father probably turned to distant family members who were senior officers in the army to use their influence to get the eighteen year old Tesla released from conscription, on the grounds that he was still extremely weak from his prolonged illness. Some sources suggest that Tesla avoided the army by the simple expedient of running away from Smiljan. In any case, Tesla did spend most of 1874 in the wild country of the Austrian mountains. Dressed as a hunter, he hiked, hunted, and fished, and continued nursing visions of the fantastic inventions he one day hoped to build. In his autobiography, Tesla claimed that it was at this age that he began to turns his thoughts seriously to invention, as the practical manifestation of all the peculiar visions he had had throughout his life.

When one speaks of "visions" with regards to Tesla, it is not a metaphor, as when one speaks of an artist or even another scientist having a "vision" for his or her creations or research. According to his own account, Tesla was subject to an immensely strange and undiagnosed physical condition throughout his life, a condition which he referred to with phrases such as "tormenting appearances", and "luminous phenomenon".

Tesla describes in his own words how the condition manifested when he was a young child:

"In my boyhood I suffered from ... the appearance of images, often accompanied by strong flashes of light, which marred the sight of real objects and interfered with my thought and action. They were pictures of things and scenes which I had really seen, never of those I imagined. When a word was spoken to me the image of the object it designated would present itself vividly to my vision and sometimes I was quite unable to distinguish whether what I saw was tangible or not. This caused me great discomfort and anxiety. None of the students of psychology or physiology whom I have consulted could ever explain satisfactorily these phenomena. They seem to have been unique although I was probably predisposed as I know that my brother experienced a similar trouble."

He goes on to explain that he believed this condition to be "a reflex action from the brain on the retina"—that is, a malfunction of his optical vision, rather than a psychological condition which would have produced hallucinations. In all other respects, he believed himself to have been a normal, "composed" child, and the fact that his brother had experience a similar condition seems to argue that its basis was physiological rather than emotional.

Because he could not control the appearance of these pictures, they were a considerable burden to Tesla as a boy. He describes the nightmarish scenario of attending a funeral during the day, only to go to bed at night and see the images of the corpse and the coffin and the mourners in black superimposed on his eyes as he lay in bed attempting to sleep.

(Touchingly, Tesla leaps from this description of a young boy's frightening experience to a conjecture as to how to a person could use that experience "to project on a screen the image of any object one conceives and make it visible. Such an advance would revolutionize all human relations. I am convinced that this wonder can and will be accomplished in time to come; I may add that I have devoted much thought to the solution of the problem." Tesla wrote this in 1919, when the film industry was still in its infancy. As with so many of Tesla's technological visions, machines that can project images onto a screen are of course a daily facet of life in the twenty first century; by the time of his death in 1943, televisions had only just been invented, and were not present in most people's homes.)

Tesla learned to exert control over these involuntary optical visions: first, by deliberately picturing something else he had seen, in effect summoning a new vision to take the place of the one that had come over him against his will. However, he could only picture things that he had seen with his own eyes, and he had to keep summoning new images in order to keep the involuntary ones at bay. Because he was only a young boy in a small village in the nineteenth century, he soon ran out of familiar objects to picture, and the more familiar the summoned pictures were, the less power they had to keep the involuntarily pictures at bay. His solution, as a twelve year old, was to begin using his imagination to create new objects, then new people, and new places. The pictures of these imaginary creations were blurry to him at first, but they slowly took on detail, until they were just as vivid to him as anything from his real life. He describes using this extraordinary power throughout his childhood to travel to new countries and make friends with the people who lived there—experiences and relationships that had as much life and color as if he had really been there and really seen them.

It was during the summer of his seventeenth year, when Tesla was traveling in the mountains to avoid the army and build his strength back up after his dangerous bout with cholera, that he first began figuring out how to use this almost magical power of seeing for something other than his own amusement, or to ward off the pictures that came to him involuntarily. After a lifetime of believing that his longing to train as an engineer would never be gratified, the prospect of the polytechnic academy was finally before him. Tesla began to think seriously about building the inventions he hoped to someday build, and in doing so he discovered that his "peculiar affliction" had gifted him with the power to visualize his projected inventions in a completely unique way.

Any ordinary person with some scientific training who attempts to design a machine or device would begin by committing his or her ideas to paper: making notes, writing equations, drafting designs. They would then begin to build a model of their machine: fashioning the parts, fitting them together, applying a power source, and observing how the machine runs, to see if any adjustments need to be made. Tesla, however, had no need to do any of this. The facility of visualization which he had honed in order to imagine every detail of made-up foreign countries and their people translated perfectly to imagining plans, models, and experiments for his inventions without touching pencil to drafting paper. He describes the process below:

"I do not rush into actual work. When I get an idea I start at once building it up in my imagination. I change the construction, make improvements and operate the device in my mind. It is absolutely immaterial to me whether I run my turbine in thought or test it in my shop. I even note if it is out of balance. There is no difference whatever, the results are the same. In this way I am able to rapidly develop and perfect a conception without touching anything. When I have gone so far as to embody in the invention every possible improvement I can think of and see no fault anywhere, I put into concrete form this final product of my brain. Invariably my device works as I conceived that it should, and the experiment comes out exactly as I planned it. In 20 years there has not been a single exception. Why should it be otherwise?"

The corollary to Tesla's remarkable powers of visualization were the intense flashes of light that also appeared in his vision from time to time—his so-called "luminous phenomenon", which he never learned how to control as he did with the pictures in his head. Sometimes they caused him great pain, especially when they were provoked by loud noises and flashing lights, such as when he attended a shooting party when he was twenty five. Most interestingly, the light phenomenon assaulted him when he first got exciting new ideas, as if the emotional excitement induced the physical phenomenon. Tesla's vision was so extraordinary that even when his eyes were closed, he saw things that other people didn't. He describes the sight that greets him whenever he closed his eyes as,

"...a background of very dark and uniform blue, not unlike the sky on a clear but starless night. In a few seconds this field becomes animated with innumerable scintillating flakes of green, arranged in several layers and advancing towards me. Then there appears, to the right, a beautiful pattern of two systems of parallel and closely spaced lines, at right angles to one another, in all sorts of colors with yellow-green and gold predominating. Immediately thereafter the lines grow brighter and the whole is thickly sprinkled with dots of twinkling light. This picture moves slowly across the field of vision and in about 10 seconds vanishes to the left, leaving behind a ground of rather unpleasant and inert grey which quickly gives way to a billowy sea of clouds, seemingly trying to mould themselves in living shapes."

Perhaps as a result of having to cope with the strange and necessarily isolating effects of his unique condition, Tesla was subject to a great many compulsions and repetitive behaviors—the combination of which would today probably result in a diagnosis of obsessive-compulsive disorder, if not other conditions. He felt compelled to perform actions in numbers divisible by three. If, for instance, he walked around a city block, he would have to do it three times, or six times, and so on. He wrote that he could not enjoy his food unless he first mentally calculated the cubic capacity of the dish that was holding it—if he could not do this, then the meal he consumed held no pleasure for him. He couldn't abide the sight of certain objects, such as women's earrings, or certain smells, such as camphor (found in mothballs), and once he began a project he felt compelled to complete it, even if he lost interest in it. (In his autobiography, he complains strenuously of how this need to complete even unpleasant tasks trapped him, when he set out to read all the works of Voltaire, and discovered that Voltaire's writings ran to hundreds of thick volumes in tiny print.)

Tesla also seems to have had a condition now known as synaesthesia, a word meaning "union of the senses". Persons with synaesthesia experience sensory impressions in response to seemingly unconnected stimuli; for instance, they might see a particular color each time they hear a specific musical note, or feel a tingling in their fingers every time they see the color blue. Many synaesthetes also strongly associate individual letters of the alphabet with specific colors. Tesla describes experiencing a bitter taste in his mouth any time he saw small squares of paper suspended in liquid, as happens in certain chemical tests.

## University Career

In 1875, at the end of his mountain hiking retreat, Tesla began his studies at the Austrian Polytechnic School in the Austrian city of Graz. His first year at the university was a period of extraordinary achievement and relative financial ease: he had received a scholarship which covered all of his tuition fees and living expenses, which meant that he could have chosen to take a fairly relaxed approach to his studies. Instead, hoping to complete two years' worth of work in only one, he worked from three in the mornings until eleven at night, without taking breaks for holidays or weekends. In the end, he passed nine exams (only four were required.) He had gone to the effort of completing two years' worth of work in one year of study on purpose as a "treat" for his parents, and he was somewhat hurt and confused when his father acted as if he were unimpressed by his accomplishments. Unbeknownst to Tesla, however, his professors had been writing to his father all year, insisting that he was in danger of dropping dead from sheer exhaustion unless he either slowed down or dropped out. Perhaps "making light" of Tesla's accomplishments was his father's way of trying to show him that he needn't work himself to the bone just to impress his family. In any event, the dean wrote to praise Tesla's abilities after his remarkable performance in the examinations, and he informed Milutin Tesla that his son, worn thin though he was, was "a star of first rank".

Unfortunately, Tesla's next two years at the university did not repeat the glories of his first. The Military Frontier—a part of Croatia which for centuries had served as the Austro-Hungarian Empire's barrier against invasion by the Ottoman Empire—was preparing to be abolished within the next six years. The scholarship that had made it possible for Tesla to attend university was issued to him by the Military Frontier, and now it too was going to be abolished, which meant that after the end of Tesla's second year, his family would have to pay for his education, which they could not afford to do.

In the course of his second year, Tesla had a now-famous encounter with the German professor of electrical machinery at the Austrian Polytechnic School, a man by the name of Poeschl; the subjects he taught included theoretical physics, and experimental physics (the field of designing machines and experiments to test the propositions of theoretical physics). Poeschl was demonstrating the use of a Gramme Machine to his students, which had just arrived from Paris. (The Gramme Machine was the first generator capable of producing enough power to supply electricity to businesses on an industrial scale.) Examining the machine, Tesla became curious how the dangerous sparking effects produced by the direct current could be managed, and he suggested to Poeschl that an alternating current be used instead. Poeschl turned to the class and remarked, "Mr. Tesla may accomplish great things, but he will never do this...it is a perpetual motion machine, an impossible idea." The exposure to the problem of alternating currents, however, would prove to be seminal for Tesla's later career.

**Life in Prague**

When Tesla's scholarship money came to an end, he attempted to raise more funds by gambling for them. He played at cards, and somewhat more successfully at billiards, but the irregular lifestyle of a semi-professional gambler was not consistent with the respectable image that the Polytechnic School required of its students, and he was dismissed, forced to leave the school without finishing his degree. Mindful of the disgrace to his family, he left Austria and did not return to Smiljan; instead, his mother scraped together enough money to send him to Prague, where there was a university. Tesla writes in his autobiography that his father wished him to go to Prague so that he could finish his degree; however, Tesla biographer Margaret Cheney relates that, according to Tesla's living descendants, there are no records of him ever having enrolled in the university. However, it is possible that he attended lectures and audited classes while continuing his independent studies in the university library, continuing to investigate the problem of alternating currents.

His gambling was becoming an increasing problem during this period of time; it placed a considerable financial burden on his family, as he alternated in winning and losing large sums of money. Tesla conflicted sharply with his father over the gambling; his father had a clergyman's narrow tolerance for such things, but his mother took a more psychologically effective tactic with him. She would give Tesla money when he was cash strapped so that he could enjoy himself at the games, telling him "The sooner you lose all we possess the better it will be. I know you will get over it." Shamed, and determined to practice self-control, Tesla ceased gambling for good immediately after his mother made this speech.

Anxious to relieve the financial strain that his mishaps at the polytechnic school and elsewhere had placed on his family, Tesla took a job in Budapest, working for the brand new American telephone exchange. However, during his time in Budapest, Tesla became extremely ill. He never received a diagnosis for the illness he suffered, but its chief symptom was an excruciating sensitivity of the senses, similar to that which is suffered by the fictional character Roderick Usher in Edgar Allen Poe's short story, *The Fall of the House of Usher*.

"My sight and hearing were always extraordinary. I could clearly discern objects in the distance when others saw no trace of them... Yet at that time I was, so to speak, stone deaf in comparison with the acuteness of my hearing while under the nervous strain. In Budapest I could hear the ticking of a watch with three rooms between me and the time-piece... The whistle of a locomotive 20 or 30 miles away made the bench or chair on which I sat vibrate so strongly that the pain was unbearable. The ground under my feet trembled continuously. I had to support my bed on rubber cushions to get any rest at all. The roaring noises from near and far often produced the effect of spoken words which would have frightened me had I not been able to resolve them into their accidental components. The sun's rays, when periodically intercepted, would cause blows of such force on my brain that they would stun me. I had to summon all my will power to pass under a bridge or other structure as I experienced a crushing pressure on the skull. In the dark I had the sense of a bat and could detect the presence of an object at a distance of 12 feet by a peculiar creepy sensation on the forehead. My pulse varied from a few to 260 beats and all the tissues of the body quivered with twitches and tremors which was perhaps the hardest to bear.

"A renowned physician who gave me daily large doses of Bromide of Potassium pronounced my malady unique and incurable."

Tesla regained his health slowly, a result for which he gave credit to the healing power of exercise. In Budapest, he had become friends with a mechanic named Anital Szigety, an amateur athlete who urged him to make as much physical exertion as he could stand if he wished to recover. Together, they took many long walks across the city, and it was on one of these walks that Tesla arrived at the solution to the problem that had been vexing him ever since the day in Professor Poeschl's classroom when he decided that the Gramme dynamo would run more efficiently with an alternating current.

According to biographer Margaret Cheney,

> "It was an entirely new system that he had conceived, not just a new motor, for Tesla had hit upon the principle of the rotating magnetic field produced by two or more alternating currents out of step with each other. By creating, in effect,

a magnetic whirlwind produced by the out-of-step currents, he had eliminated both the need for a commutator (the device used for reversing the direction of an electric current) and for brushes providing passage for the current. He had refuted Professor Poeschl."

Tesla understood the game-changing nature of his new invention, but it would be another matter to get other people to recognize it. No matter how perfectly and completely Tesla could envision his induction motor, he could not build it until he persuaded someone to give him the money to do so, and it would be tricky to convince a financial backer that he knew what he was doing without a prototype to show them. On the other hand, these practical considerations almost didn't matter to Tesla: what mattered was that he had achieved his life's dream, or at least the beginning of it. All of his life, he had wanted to feel that he could call himself inventor, someone who came up with machine that would bring good to the world. Now that he had finally come up with the plan for a viable, important machine, he felt at last worthy of an inventor's name. The fact that he was still poor, hardly getting by on the salary he drew from the telephone exchange, and an ocean away from the people he needed to take an interest in his invention hardly mattered to him.

Tesla's autobiography describes the next several months as one of the happiest periods in his life. He spent the days happily lost in his head, daydreaming about (that is, meticulously visualizing designs for) all the wonderful new machines he would some day build.

# Chapter Two: The Struggle for Recognition

"When natural inclination develops into passionate desire, one advances towards his goal in seven-league boots."

> Nikola Tesla, *My Inventions*

**Paris**

Immersed though he was in the imaginary plans for his own fantastic scientific creations, Tesla apparently had sufficient time left over to apply some of his design prowess to the job he was actually being paid for. By this point, he was working in a telegraph office, where he improved the efficiency of the central-exchange apparatus. He did not receive a promotion or a higher salary as a reward for his ingenuity, but he impressed his employers (who conveniently also happened to be family friends) so much that he was offered a job in Paris, working for the phone company founded by Thomas Edison—the most famous of nineteenth century American inventors, and the man who was to have a more significant impact on Tesla's life than almost any other, for good and for ill.

Tesla was delighted to accept the job, and to begin a new life in Paris. The city delighted him, even though he was just as poor as he had been in Budapest, if not more so. Describing the straitened circumstances of his life in France, Tesla wrote,

> "The attractions were many and irresistible, but, alas, the income was spent as soon as received. When Mr. Puskas asked me how I was getting along in the new sphere, I described the situation accurately in the statement that 'the last 29 days of the month are the toughest!'"

But Tesla was able to make friends with both his French and American coworkers—the Americans with whom he worked were especially enthusiastic about spending time with him when they discovered that he was a near professional quality billiards player. He rose very early in the mornings, swam twenty seven laps in a pool (again, any repetitive task Tesla performed had to be done in numbers divisible by three), then walked for an hour to the telephone company to eat breakfast and begin work. His immediate supervisor was a man by the name of Charles Batchelor, and much to Tesla's delight, he happened to be an intimate personal friend of Edison's.

Tesla's motivation for accepting the job in Paris was primarily a desire to make contact with people who would understand and appreciate the superiority of his alternating current device compared to the direct current motors which

Edison was working with, and which were being used throughout the world. Tesla was deeply dismayed, therefore, to discover that after many of his own repeated failed experiments with alternating currents, Edison was sick of the very name of them, and was not interested in hearing any more theories about how they could be made to work. Tesla could not understand this at all. To him, it was blazingly apparent that his machine would work, and work better than anything of the kind the world had ever seen; it baffled him that others were not willing or able to look at his designs and arrive at the same conclusion.

Because of his mechanical expertise and his ability to speak German (though he was a Serb growing up in Croatia, Tesla's schools had largely taught lessons in German, and he was fluent in about eight languages), the Edison company in Paris sent Tesla to perform repairs at other plants around France and in Germany. By bringing machine parts with him from Paris, Tesla managed to construct rough prototypes that incorporated at least some of the design elements from his new invention, and use them to solve the electrical and energy problems the plants faced.

Tesla pulled off a considerable feat on behalf of the Edison company in the German town of Strassburg, where he had been sent to repair a railway station lighting plant that had been constructed for the Germany government. The Germans were refusing to pay for the plant because, during the grand opening, a small explosion caused by a short circuit had blown out a portion of a wall—right before the eyes of Kaiser Wilhelm I, the first ruler of unified Germany, a man no one wished to disappoint. Tesla's German speaking skills and his technical expertise went so far towards winning the Germans over that he became good friends with Strassburg's mayor, who even attempted to round up investors for his alternating current device, though without success. (He did share with Tesla a portion of an extremely rare wine, which he had buried in the ground many years before—drinking it was such a significant experience, apparently, that Tesla devotes some paragraphs to it in his autobiography.)

For the feat Tesla had accomplished in Strassburg, which saved the Edison company a great deal of money, he had been promised a large bonus on his return to Paris. But each of his three immediate superiors declared that it was the responsibility of one of the others to produce the money, and the result was that it was never paid. Tesla was furious; he needed the money to build his invention, and he was beginning to be afraid he would never get it. Furious at this dishonest treatment, Tesla resigned from the Edison telephone exchange in Paris.

His supervisor, Charles Batchelor, had figured out a long time ago that Tesla was a genius of the first water, with no equal in the world of electrical engineering unless it was Edison himself. So he gave Tesla the best advice and help that it was in his power to give: he urged the young inventor to go to America to try his fortunes, and he gave Tesla a letter of introduction to Thomas Edison to help him do so.

Tesla made immediate arrangements for travel to the United States. He set about his transatlantic journey with a unique sense of determination. On his way to the train station, mere moments before the train pulled away, he discovered that he had lost, or been pickpocketed of, his ticket and almost all of his money. He had to run alongside the train and swing himself aboard just as it was picking up speed; once inside, he scraped together the pennies for a cheap ticket. When he reached the

*Saturnia,* the ship that would take him across the ocean, he was able to persuade a porter to let him have the berth he had reserved, though he could not prove that he had paid for it. Tesla writes that he spent most of his ocean voyage seated near the stern of the boat, watching in case anyone should fall overboard: he, with his strong swimming ability, would undoubtedly be the one to save them. (He also writes that when he was older, and possessed more common sense, he shuddered to remember how reckless he had been.)

When he arrived in Manhattan weeks later, he had nothing with him but a sheaf of personal papers—articles he had written, illustrations, notes for his inventions—and a few coins. It was a story that would be told over and over again during this era of American history, the penniless European immigrant in America, hoping to try his luck and build his fortune.

Tesla, it would turn out, was luckier, and more successful, than most of them.

## Manhattan

Tesla's arrival in the United States was not exactly the stuff of fairy tales, but he was in a far better situation than most of his fellow immigrants arriving on the ship with him. His clothing indicated that his station in life was that of an educated, professional man, which meant that the customs officials treated him with a bit more respect than the transplanted shepherds and farmers he arrived with. Most of the *Saturnia*'s other passengers would be hired into work gangs, where they would slave for low wages over long grueling work days. Tesla, on the other hand, had arrived rich in contacts, if not rich in money. And the opportunities for which America was so famous seemed to positively leap out Tesla from the moment he disembarked the ship. Walking down the street, he encountered a man in a shop window having difficulty with a bit of broken machinery. Tesla stopped and offered to repair it for him, and the man, astonished by the providential arrival of this strange young foreigner with the magic touch, paid him twenty dollars. This must have been greatly welcome to Tesla, who had arrived in the United States with only four cents in his pocket.

Tesla had this to say about his early experiences in the United States:

"I wish that I could put in words my first impressions of this country. In the Arabian Tales I read how genii transported people into a land of dreams to live through delightful adventures. My case was just the reverse. The genii had carried me from a world of dreams into one of realities. What I had left was beautiful, artistic and fascinating in every way; what I saw here was machined, rough and unattractive. A burly policeman was twirling his stick which looked to me as big as a log. I approached him politely with the request to direct me. "Six blocks down, then to the left," he said, with murder in his eyes. "Is this America?" I asked myself in painful surprise. "It is a century behind Europe in civilization." When I went abroad in 1889 - 5 years having elapsed since my arrival here - I became convinced that it was more than one hundred years AHEAD of Europe and nothing has happened to this day to change my opinion."

There is no telling at precisely what point Tesla changed his mind about the American nation being hopelessly backwards, but it is just possible that it coincided with his long-anticipated first meeting with Thomas Edison.

## Meeting Thomas Edison

According to Tesla biographer Margaret Cheney, Edison was having a very difficult day when Tesla walked into his office for the first time. New York was the first city in the United States to have widespread electrical lighting, and Edison's companies were responsible for all of it. But the art of electrical engineering was still in its infancy, and accidents happened constantly. Fires broke out; wires snapped; horses got electrified and startled into a panic just walking down streets where the electric wiring had been set up, because they were absorbing electricity through their metal shoes. And Edison did not employ nearly enough trained engineers to deal with the issues that were cropping up; he was constantly receiving phone calls demanding that he send out a man to make a repair or an installation or restart a dynamo, and there simply weren't any.

Unlike Tesla, who, since boyhood, had cherished dreams of inventing machines that would benefit all of humanity, and whose chief joy in inventing was in the somewhat artistic process of coming up with ideas for machines and visualizing how they would work, Edison was in a much more practical business: the business of inventing things for money. He had not attended a premier polytechnic university; he was, like most Americans who came to prominence in a century that venerated the frontiersman and the log cabin born self-made man, a self-taught electrical engineer and businessman. Tesla, however, respected Edison all the more for the fact that he had not had the advantages of early scientific training and education, and had spread the incandescent light bulb across the world nonetheless.

Tesla would later write that the day he met Thomas Edison was the most exciting and profoundly moving experiences of his life to that point. Batchelor's letter of introduction, which Tesla had brought with him from Paris, emboldened him to walk directly into Edison's office and attempt to introduce himself. Harried, but curious, Edison read the letter from Batchelor that Tesla presented him with: "I know two great men and you are one of them," Batchelor had written. "The other is this young man." Cautiously impressed, Edison asked that Tesla explain his qualifications.

It should come as no surprise that Tesla immediately launched into a description of his design for the alternating current induction motor; but he found that what his superiors in Paris had told him about Edison being unwilling to hear a word spoken on the subject of alternating current was nothing more than the truth. However, Edison was, as we have discussed, just at that moment in desperate need of a trained engineer. There was a great demand for electrical lighting on ships, because of the gorgeous spectacle they made when traveling down the water at night with all lights burning—this, despite the fact that all electrical systems were then fire prone, and nothing could be more dangerous or devastating than a ship catching on fire when it was at sea. Lighting on sea-going vessels had to be approached with the utmost caution, and one ship, the S.S. *Oregon*, had been delayed in the docks for several days past its expected launch date because the lighting plant on the ship was broken. The crew had phoned that morning to demand that Edison send them an engineer; he had had no such engineer when he received the phone call. But once Tesla had walked into his office and made his pitch, Edison realized he had just the man he needed. Delighted with the opportunity to prove himself, Tesla went directly to the ship and began work on repairs.

Tesla himself describes the reaction that Edison had when he discovered how quickly Tesla was able to solve the problem on the ship:

"At 5:00 a.m. when passing along Fifth Avenue on my way to the shop, I met Edison with Batchelor and a few others as they were returning home to retire. 'Here is our Parisian running around at night,' he said. When I told him that I was coming from the Oregon and had repaired both machines, he looked at me in silence and walked away without another word. But when he had gone some distance I heard him remark: 'Batchelor, this is a damn good man.'"

The professional working relationship between Tesla and Edison was bound to be short-lived, however. Cheney describes the gulf of differences between their personalities, philosophies, and general approaches to life and business. Simply put, Edison lacked Tesla's training and his culture; he was the embodiment of a certain rough-hewn American archetype that disdained the trappings of too much civilization, and even too much education of the theoretical variety. Long after parting ways with Edison, Tesla remarked that he had often watched Edison labor at experiments with a kind of helpless pity; Tesla's formal education in the sciences had taught him that the ability to make calculations could save a great deal of labor in the experimental front. That is to say, Tesla could work out on paper (or more likely, in his head) whether or not a machine would work before he ever built the prototype. Edison was suspicious of this invisible style of tinkering, however, and refused to take advantage of Tesla's experience to work around his own limitations.

There was also the fact that Edison was intelligent enough to realize that Tesla's alternating current system would probably work, that it would work better and more safely than direct current motors, and that if Tesla ever got his system fully developed and integrated into a sound business model, it would probably render Edison's electrical engines virtually obsolete. When Edison had first started in the electric lights business, the gas companies had opposed him at every turn, sensing the he was out to replace them; now Edison was finding himself in the same position with Tesla's alternating current induction motor.

In any event, Tesla ceased working for Edison once it became apparent that Edison was only going to make use of Tesla's abilities to earn money for his own business, and had no intention of assisting Tesla in a way that might lead to his becoming serious competition for Edison. The final blow came when Tesla offered to repair and redesign the electric dynamos in Brooklyn that ran its electric powered trolleys. It was an absolutely immense undertaking, but Tesla promised Edison that once his redesigns had been implemented, Edison would start saving considerable money.

Edison told Tesla that if he could actually pull it off, there would be fifty thousand dollars in it for him. Perhaps Edison never expected Tesla to manage it; or perhaps, as he later claimed, he was only joking, and the European Tesla failed to grasp the nuances of American humor. Either way, Tesla worked virtually without stopping for an entire year, until the project was completed. He was crushed when Edison told him there was no fifty thousand dollars waiting for him. Edison attempted to mollify Tesla somewhat by offering to raise his eighteen dollar an hour pay rate to twenty eight dollars an hour, a huge hourly salary by turn of the century standards. But Tesla was bitterly reminded of the bonus that had been promised to him by Edison's managers in Paris, when he had repaired the lighting plant at the Strassburg railway station; he had never received that money either.

It was one thing for the managers of Edison's phone exchange across the ocean to lie to and cheat him, and quite another thing for the much-admired Edison to do it. Tesla resigned; Edison told him that he was making a mistake, but there was never any real possibility of their working together successfully. Tesla was too brilliant to remain anyone's employee for long, and especially to remain in the employ of another genius whose egotism would not permit him to give the younger man a fair deal.

## The Tesla Electric Light Company

One of the factors that made Tesla's resignation a risky move was the Panic of 1884, which was in full swing when Tesla left off working for Edison. It was a bad year for small businesses, speculation, investors, and entrepreneurs. But Tesla was a unique figure and a unique talent, and he hadn't been unemployed long before he was approached by investors who had heard of his work for the Edison company and his ideas for the induction motor. Indeed, Tesla scarcely ever stopped talking about the induction motor, believing that sooner or later the right people would hear about it and would help him build it. He was proven correct; no sooner was he out from behind Edison's shadow than he met a group of wealthy venture capitalists who wanted to give him his own company, under his own name, where no jealous rivals or employers could stop him from finally fully developing his induction motor.

First, however, his investors insisted on a more prosaic application of his abilities: arc lamps, in which light is generated by an electric arc, were applicable to many industrial uses, and were favored for street lighting. To satisfy them, Tesla came up with his own design for an improved arc lamp that was safer and used power more efficiently than the standard model, and got one of his first commercial patents as a result. The town of Rahway, New Jersey, which hosted the headquarters of the Tesla Electric Light Company, was the first to adopt them. However, once Tesla had provided the patent, he was paid for them in stock shares (which would have to mature for some time before they were worth anything) and effectively dismissed, without having ever been given a chance to work on the induction motor. Dismayed and disheartened, Tesla left the company and looked, again, for a new opportunity.

**Tesla Electric Company**

For a year following his departure from his Electric Light company, Tesla was forced to make ends meet however he could. Because of the financial depression, the only work he could find from 1886 to 1887 was on one of the same physically ruinous labor gangs that so many of his fellow immigrants from the *Saturnia* had been hired into. (Ironically, the same multi-millionaires who were building their fortunes on the backs of laborers such as Tesla would soon be clamoring to invest in Tesla's genius.) Luckily, the foreman of one of these work gangs seemed to recognize that Tesla's talents were being wasted at hard labor, and he arranged for Tesla to meet a manager from America's most famous telegraph company, Western Union. This manager took a strong interest in Tesla's work on the induction motor, and with his help, Tesla formed another company, this time specifically geared to develop the invention Tesla had visualized into existence in that glorious moment back in Budapest over five years before.

# Chapter Three: AC versus DC

"He began laboring like one of his own dynamos, day and night without rest. Because it was all there in his mind, he needed only a few months to start filing patent applications for the entire polyphaser AC system. This was in fact three complete systems for single-phase, two-phase, and three phase currents. He experimented with other kinds too. And for each type he produced the necessary dynamos, motors, transformers, and automatic controls."

Margaret Cheney, *Tesla: Man Out of Time*

Perhaps the only moment of his life when Tesla was as happy as when he figured out how to make his induction motor work was the moment when he was finally given the freedom to fabricate, patent, manufacture, and release his invention to the world, where it could at last benefit humanity in the way it was meant to.

There was no uniformity in the electrical engineering infrastructure of the United States at this time. A dominant system had yet to emerge; the various companies providing electric lights were all in competition with one another to spread the usage of the specific invention that their own lighting systems were built upon. And each company had developed their lighting to fulfill a specific market niche; for instance, some company's lighting systems were particularly suited to lighting factories, but not to providing safe electric lights for houses, and vice versa.

One company which was emerging as the lead competitor for Thomas Edison and the Edison Electric Company was the company belonging to George Westinghouse; he was an early adopter of alternating current systems, which made him a natural ally of Tesla's. And Tesla was already beginning to make a name for himself: in the year 1891 alone, he was granted forty separate commercial patents. When word of this began to circulate, the engineering world finally began beating a path to Tesla's door. Westinghouse, who realized almost immediately that Tesla's alternating current induction motor was going to change the direction in which technology was developing, came to visit Tesla personally and offer to license his patents from him. Tesla received a lump sum of cash, and a number of valuable shares, and was able to forget about any money troubles for a short time. Additionally, Westinghouse hired him to consult for his company in Pittsburgh for a salary of $2000 a month.

For neither the first nor the last time in his life, Tesla faced the problems of jealous inferior intellects impeding his progress. Westinghouse's engineers balked at building the motors precisely to his specifications because they wished to use already existing machinery, rather than built every part from scratch. According to Margaret Cheney, they wasted a great deal of time and money attempting to prove that Tesla's motor could run according to their projections, only to be forced to give up and build the motor that Tesla had designed, which, of course, ran perfectly. Because of this, the transition process of outfitting Westinghouse to make induction motors ran far over projected costs; eventually, it would force Westinghouse to make a merger that had severe consequences for Tesla's financial future.

## Return Visit to Croatia

Much to Tesla's pride and delight, he was naturalized as an American citizen on July 30, 1891, about five years after his arrival in the country. Not long afterwards, however, he was sufficiently exhausted by his work in America to feel that his health was in some danger, and that he badly needed a rest. So in September of 1891, he traveled to Paris to attend the International Exposition, a sort of world's fair displaying technological advances from all over Europe. From there, he traveled home, to Croatia, where he saw his widowed mother and his sisters again for the first time since he was forced to leave the Polytechnic School in Austria.

Tesla gives this account of his further travels in Europe:

"I went to London where I delivered a lecture before the Institution of Electrical Engineers. It had been my intention to leave immediately for Paris in compliance with a similar obligation, but Sir James Dewar insisted on my appearing before the Royal Institution. I was a man of firm resolve but succumbed easily to the forceful arguments of the great Scotsman. He pushed me into a chair and poured out half a glass of a wonderful brown fluid which sparkled in all sorts of iridescent colors and tasted like nectar. "Now," said he, "you are sitting in Faraday's chair and you are enjoying whiskey he used to drink." In both aspects it was an enviable experience. The next evening I gave a demonstration before that Institution, at the termination of which Lord Rayleigh addressed the audience and his generous words gave me the first start in these endeavors. I fled from London and later from Paris to escape favors showered upon me, and journeyed to my home where I passed through a most painful ordeal and illness. Upon regaining my health I began to formulate plans for the resumption of work in America."


The English scientist Michael Faraday had been the last great mind in the field of electrical engineering to come along before Tesla, and therefore it is to be assumed that Sir James intended by this behavior to confer an honor upon Tesla.

Tesla was in search of his next big idea—the flash of inspiration that would inspire his next world changing invention. As it happened, the moment of inspiration struck during his restorative European sojourn. In his autobiography, Tesla describes hiking through the mountains and getting caught in a thunderstorm. At first, there was no rain, only claps of thunder; but then there came a lightning strike, and a torrential downpour of rain followed. Tesla was struck by the fact that the rain had changed its behavior in response to the electrical excitement produced by the lightning, and from this observation he derived any number of fascinating ideas:

"Here was a stupendous possibility of achievement. If we could produce electric effects of the required quality, this whole planet and the conditions of existence on it could be transformed. The sun raises the water of the oceans and winds drive it to distant regions where it remains in a state of most delicate balance. If it were in our power to upset it when and wherever desired, this mighty life-sustaining stream could be at will controlled. We could irrigate arid deserts, create lakes and rivers and provide motive power in unlimited amounts. This would be the most efficient way of harnessing the sun to the uses of man. The consummation depended on our ability to develop electric forces of the order of those in nature. It seemed a hopeless undertaking, but I made up my mind to try it and immediately on my return to the United States, in the Summer of 1892, work was begun which was to me all the more attractive, because a means of the same kind was necessary for the successful transmission of energy without wires."

Tesla would return to the United States full of energy and inspiration for this potentially world changing project. But in the mean time, he had other problems.

**Edison's revenge**

At the beginning of his career, Thomas Edison and his electric lighting systems had faced powerful opposition from the gas companies; gas lighting had been the standard powered lighting system at use in homes, business, and public facilities for decades, and electric lights threatened to make them obsolete. In order to get the important public contracts he needed to make electric lights replace gas lights as the national standard, Edison threw as much energy into public relations and propaganda as he did into inventing and engineering. The safety issues involved with gas were real and dangerous, and Edison played them up before the public eye, trying to convince the public that they risked fires, explosions, and deaths if they did not switch from gas lighting to electric lighting.

When Edison learned that his former employee, Tesla, had gone to work for George Westinghouse's company, and was on the verge of releasing safe, efficient alternating current induction motors that would threaten the supremacy of Edison's direct current system, he took it as a personal insult. He immediately began drumming up a propaganda storm against alternating currents, attempting to convince the public they were unsafe. The problem, of course, was that they were no such thing; Tesla's entire motivation in developing the alternating current motor was to make it safer than the direct current motor he had witnessed sparking dangerously during the demonstration of the Gramme dynamo back at in Professor Poeschl's classroom at the Austrian Polytechnic School.

But to Edison, the facts, in this case at least, scarcely mattered. He was invested in direct current and adamantly opposed to alternating current, and he would do whatever he must to turn the public against alternating current. In Margaret Cheney's words, "accidents caused by AC must, if they could not be found, be manufactured, and the public alerted to the hazards. Not only were fortunes at stake in the War of the Currents but also the personal pride of an egocentric genius."

Because of the almost immediate success of Tesla's induction engine, rival electrical companies were forced to buy licenses and use his designs, which were patented, or challenge the patents—either legally, by claiming that their own inventions and patents pre-dated his, or commercially, by coming up with a design for an engine that was like Tesla's, but also different enough to qualify for a new patent. All of Tesla's patents were upheld in court, however, and despite Edison's best efforts, Tesla's design dominated the market almost completely.

Edison was undeterred, however, and his methods were not only dishonest, they were shockingly cruel. In West Orange, New Jersey, "Edison was paying schoolboys twenty-five cents a head for dogs and cats, which he then electrocuted in deliberately crude experiments with alternating current. At the same time he issued scare leaflets with the word 'WARNING!' in red letters at the top." Even more outrageously, Edison told people that George Westinghouse was running a disinformation campaign against *him*. Westinghouse, however, was reluctant to sink to Edison's level, even to correct the disinformation being issued against him. He was more interested in winning the International Niagara Commission's bid to devise a system that would harness the power of the Niagara Falls.

One note about Edison's electrocution of animals (and a warning for graphic descriptions of extreme animal cruelty): in 1903, a female Asian circus elephant named "Topsy" was killed by alternating current at a public exhibition at Coney Island, in New York. Topsy had been imported to the United States when she was a calf, but her owners had falsely publicized her as the first Asian elephant born in captivity in America. She had been deemed a "dangerous" elephant, owing to accidents that occurred with her handlers, probably in reaction to maltreatment and abuse, and after being passed from owner to owner, it was decided that she must be killed, and that her owners would make money off her death by turning her execution into a public spectacle. Of course, there were difficulties inherent in killing such a massive animal. Originally, her handlers intended to strangle her slowly by wrapping rope around her neck and hanging her from a crane, but it was feared that this would not kill her at all, or at least not quickly enough. To make sure that she died, she was also fed poison, and in the showiest part of the spectacle, electrocuted—with, of course, alternating current. The killing of Topsy the elephant has been touted as a shocking piece of anti-Edison lore for decades; it even featured in an episode of the American cartoon *The Simpsons*. But whatever other acts of animal cruelty Edison was guilty of, and there were many, Topsy's death was not his doing. The so called War of the Currents was long over by the time of Topsy's killing in 1903, and Edison himself was not present in Coney Island when she died. This is merely one of the common myths that people tell to illustrate the intensity of Edison's obsession with preventing alternating current from becoming the American standard.

Edison may not have killed Topsy the elephant, but in his absolute determination to make the name of "Westinghouse" synonymous with danger and death, he did end up contributing to the electrocution of a human being. It was as a result of direct lobbying by Edison that electrocution came to replace hanging as the primary means by which American prisoners sentenced to death were executed. (In fact, the word "electrocute", which is sometimes used erroneously to mean "electrify", means "to be killed by electricity". It is a portmanteau of the words "electrify" and "execute".) Once the announcement of the change from hanging to electrocution had been made, Edison made certain that everyone knew that the electrocutions would be carried using alternating current and the Westinghouse patents.

And the War of the Currents went on from there. Edison was giving public demonstrations of the lethal capabilities of alternating current by throwing dogs, cats, and occasionally livestock onto electrified metal platforms and killing them in front of spectators. However, karma, so to speak, in the form of financial difficulties, was soon to catch up with him. Edison's business enterprises were too far flung; Edison General Electric was soon forced to form a merger with the second of the three big electrical companies in America, Thomson-Houston, in order to survive. (The third major electrical company, Westinghouse, held out against merging for some time.) The new company that emerged, of which Edison was not president, was called General Electric—the same company which famously manufactures most of America's lightbulbs today.

## Westinghouse Electric and Manufacturing Company

The underhanded campaign that Edison had launched against George Westinghouse's company was paying off. Westinghouse's stock value was beginning to fall. This was also due to problems that Westinghouse was facing internally: he had been forced to undertake the enormous expense of converting all of his machinery to the Tesla polyphasic model, and the initial cash expenditure had not paid off yet. Capitulating to pressure from the bankers who held his loans, Westinghouse began to arrange a merger: not with General Electric, but with a number of smaller electric companies, such as U.S. Electric and Consolidated Electric Light.

There was one insurmountable problem standing in the way of the merger that would create the Westinghouse Electric and Manufacturing Company, however. The terms of George Westinghouse's original arrangement with Tesla for the use of his inventions included substantial royalty payments, which Tesla had not yet received, but which he had an unshakeable legal claim to. This effectively placed the fate of the Westinghouse company in Tesla's hands. George Westinghouse could not cancel the royalties contract; the only possibility of his getting out from under the liability was for George

Westinghouse to appeal personally to Tesla, and ask him to tear the contract up.

One of the qualities that Tesla is best remembered for now is his general inattention to practical financial matters. He had what one might call an artistic or poetic soul; his vision for his inventions was one of transformation for the world, not one of gaining vast quantities of personal wealth and power. He wanted enough money to live well and fund his experiments, but unlike Edison, he was not a businessman and was not interested in power over the marketplace. To Tesla, what was most important was that his alternating current system be adapted throughout the United States, and eventually the world.

The fact was, Tesla was probably not capable of appreciating exactly how much money he was really entitled to; he enjoyed having money and spending it on personal comforts, but the kind of financial concerns that translate to power in the business world were uninteresting to him. In the currency of the 1890's, Tesla was legally owed about twelve million dollars in royalties; in 2016 currency, this figure would amount to more than three hundred million dollars.

George Westinghouse had been a friend to Tesla; he had given Tesla the opportunity he dreamed of to bring his alternating current motor into existence and show its usefulness to the world. And now Westinghouse was telling him, "Your decision determines the fate of the Westinghouse Company." If Tesla renounced his royalties, the company could continue with the merger and continue to spread the gospel of alternating current power. Even if Tesla did not do so, his fate was still uncertain; he would have to deal with bankers over money matters that he did not fully understand. Since Tesla did not fully grasp the financial aspect of the situation, he let his personal feelings be his guide. He told Westinghouse that he would tear up his old contract. In so doing, he forfeited not only the money he had already earned and not been paid, but untold millions of dollars that would have accrued to him in the future. Instead, he sold his patents outright to Westinghouse for a lump sum of cash, about sixty thousand dollars—a tiny fraction of the money he was entitled to. When Tesla was a little older, and facing serious cash shortages, he must have thought back on this act of astonishing generosity with some wistfulness, at the very least.

# Chapter Four: Dreams and Visions

## The Most Famous Scientist in the World

Tesla's chief joy in life was puttering around in his private laboratories, inventing things, bending electricity to his will, and demonstrating the power of his inventions to his admiring and interested friends (and wealthy potential investors). For about a decade after the sale of his patents to George Westinghouse, he had enough money to keep him comfortably insulated from the necessity of turning these fantastical, almost magical inventions into salable investments.

For several years in the early 1890's, Tesla gave a series of lectures to the public. He traveled around the world to give talks on the powers of electricity and give demonstrations of his inventions that baffled the imaginations of all who witnessed them.

In February of 1892, Tesla gave a lecture in London entitled "Experiments with Alternate Currents of High Potential and High Frequency" to the Institution of Electrical Engineers and the Royal Institution of Great Britain. Two weeks later, he traveled to Paris, where lectured on "Experiments with Alternate Currents of High Potential and High Frequency" before the Societe Francaise de Physique.

When giving these lectures, Tesla encountered a fundamental problem of language. He spoke many languages, but the one he needed most, the language of science, had not been invented—or at least, it had not developed to include the terminology he needed to make his discoveries, and the phenomena he was displaying to the crowds, intelligible to researchers today. He was, simply put, so far ahead of his time that there was no context for his discoveries. There was nothing unscientific about them; every experiment he conducted had been duplicated dozens of times to the most rigorous standards of the scientific method. But Tesla had to invent his own vocabulary in order to talk about them. He gave names to equipment and the effects they produced according to his own rather poetic perceptions of their appearances; other times, he named them after people, such as the Serbian poets whose national epics he had committed to memory in childhood.

The best efforts of contemporary scientists have translated only a portion of Tesla's demonstrated findings into concepts that are understood today. For some things, we still do not have the proper vocabulary. The scientific world is only just beginning to catch up to Tesla, and in some respects he is still beyond us.

In the following excerpt from Tesla's London lecture to the Institute of Electrical Engineers, one can see how, in an effort to communicate his findings to a room full of people who could follow them only in part, he relied on verbal descriptions of the optical phenomenon he produced, and the actions he took with the displayed machinery to produce them:

"Here is a simple glass tube from which the air has been partially exhausted. I take hold of it; I bring my body in contact with a wire conveying alternating currents of high potential, and the tube in my hand is brilliantly lighted. In whatever position I may put it, wherever I may move it in space, as far as I can reach, its soft, pleasing light persists with undiminished brightness.

"Here is an exhausted bulb suspended from a single wire. Standing on an insulated support. I grasp it, and a platinum button mounted in it is brought to vivid incandescence.

"Here, attached to a leading wire, is another bulb, which, as I touch its metallic socket, is filled with magnificent colors of phosphorescent light.

"Here still another, which by my fingers' touch casts a shadow—the Crookes shadow, of the stem inside of it.

"Here, again, insulated as I stand on this platform, I bring my body in contact with one of the terminals of the secondary of this induction coil—with the end of a wire many miles long—and you see streams of light break forth from its distant end, which is set in violent vibration..."

Tesla biographer Margaret Cheney believes that scientists Frédéric and Irene Joliot-Curie, Henri Becquerel, Robert A. Millikan, Arthur H. Compton, Ernest Orlando Lawrence, and Victor F. Hess, all of whom won Nobel prizes, took inspiration from or built upon work that had been begun by Tesla, who received many honors, prizes, and medals in his lifetime, but never won a Nobel himself. His ability to inspire the creativity of his fellow scientists may have led to even more scientific breakthroughs than he himself discovered.

**The Death of Duka Mandíc**

Tesla's lecture series made him extremely famous; not merely by the standards of scientists, but by any standards. He became a world famous celebrity in about four months' time, and was prepared to extend his lecture tour even longer. Unfortunately, personal tragedy struck Tesla's family just a few months after he began traveling Europe. In April of 1892, his mother was stricken with a sudden, serious illness. But the way that Tesla found out about her illness is one of the most remarkable stories he has to tell in his autobiography.

At that point in his life, Tesla had not seen his mother, or anyone else in his family, for several years. He was not only in the midst of his lecture tour; prior to going abroad, he had spent the entire year since leaving the Westinghouse company engaged in the kind of experimental personal research that delighted him so much. Most of his research during that period was in the field of radio waves. He had noticed a particular reaction produced by his grounded transmitter, which sent an electrical current through the earth. The potential applications to wireless communication fascinated him, and he had become obsessed with studying the problem, to the point of working for more than a year at the same grueling pace that had so frightened his professors at the Polytechnic School and had so impressed Thomas Edison.

Tesla enjoyed excellent health the majority of the time. As a man in his sixties, he bragged that he had never lost nor gained a single pound since he was an adult, and that the suits he wore were made to the same measurements and specifications that had been taken when he was in his early twenties. He exercised vigorously, and because of all this he was able to sustain a pattern of twenty hour working days for an impressively long period of time. Sooner or later, however, a "reaction" would come over him. His health would begin to fail him, he would become temporarily disabled. Often, he would suffer strange attacks that baffled diagnosis or treatment by doctors. (Indeed, by his early thirties, Tesla had virtually given up on doctors altogether.)

In 1892, the peculiar symptoms which he suffered after working for more than a year on the grounded radio transmitter were not problems of a kind that any doctor would understand, unless they were intimately familiar with Tesla's medical history. Tesla partially lost his ability to visualize—to call images from his life and his imagination before his eyes with as much vividness as if they really lay before him.

This idiosyncratic optical phenomenon, which had plagued Tesla as a child until he learned how to control it for his own use, was integral to the way Tesla worked. He did not have to draft his machines, or build endless prototypes; he simply visualized them, and they would work. But one day, after his year of nonstop work finally caught up with him, he fell into a very deep sleep. Previously, he had been able to visualize anything and everything he had ever seen throughout his entire life in the most excruciating detail. But when he woke up, the only images he could visualize in this way were images from his early childhood. Naturally, images of his mother were a central feature in all these early childhood impressions. As he describes it,

"Night after night, when retiring, I would think of them and more and more of my previous existence was revealed. The image of my mother was always the principal figure in the spectacle that slowly unfolded, and a consuming desire to see her again gradually took possession of me. This feeling grew so strong that I resolved to drop all work and satisfy my longing. But I found it too hard to break away from the laboratory, and several months elapsed during which I had succeeded in reviving all the impressions of my past life up to the spring of 1892."

Tesla had been receiving letters from Croatia for a few months leading up to this period, letters which indicated that his mother's health was worsening. He could not tear himself away from his experiments at first, but as he began receiving letters and invitations to give lectures and receive prizes from around the world (the induction motor was well on its way to revolutionizing the industry, and the world had taken notice of its inventor) he decided that he should accept invitations in London and Paris, and immediately thereafter travel to visit his family.

The urgency he felt surrounding this visit home was compounded by the fact that as he regained his ability to visualize things from the rest of his life, he had a vision of kinds—a visualization that he had not intentionally created, but which appeared before his eyes involuntarily, prognosticating future sorrow.

"In the next picture that came out of the mist of oblivion, I saw myself at the Hotel de la Paix in Paris just coming to from one of my peculiar sleeping spells, which had been caused by prolonged exertion of the brain. Imagine the pain and distress I felt when it flashed upon my mind that a dispatch was handed to me at that very moment bearing the sad news that my mother was dying. I remembered how I made the long journey home without an hour of rest and how she passed away after weeks of agony! It was especially remarkable that during all this period of partially obliterated memory I was fully alive to everything touching on the subject of my research. I could recall the smallest details and the least significant observations in my experiments and even recite pages of text and complex mathematical formulae."

Tesla speaks of "remembering", but that is merely a figure of speech; the event had not actually happened, and he was not confused on that point. Months later, however, during his 1892 lecture tour, Tesla was returning to the Hotel de la Paix after finishing his lecture to the Societe Francaise de Physique, when he was handed a telegram informing him that his mother was on her deathbed—not exactly as he had seen it in his vision, but nearly enough. Tesla was exhausted; upon bidding farewell to Sir William Crookes after giving his London lecture, Sir William had written to him, telling him that he looked like he was close to a nervous collapse from overwork, and that he should retreat to the mountains of his native country without even taking the time to reply to the letter in his hands. For once, Tesla was prepared to acknowledge that he was over-exerting himself, but he had the Paris lecture to get through first. When the telegram from his family arrived, he was on the point of catatonia; but he summoned the necessary strength to race for the train station, catching a train to Croatia just as it was leaving.

Luckily, Tesla reached his mother in time to speak with her and sit with her for a few hours. Her first words to him were also his last: "You've arrived, Nidžo, my pride." But eventually, he was forced to go to a hotel and rest, and his mother died that night. Remarkably, Tesla had already convinced himself that if his mother were to die while he was not by her side, she would appear to him one last time. In England, during his recent visit with Sir William Crookes, there had been a discussion of supernatural phenomenon and spiritualism. In the late nineteenth century, the spiritualist movement had swept the upper classes and even the intelligentsia of Europe and America, with people holding séances in darkened parlors, attempting to communicate with the dead. Tesla appeared to believe that the soul might conceivably manifest after death as some form of electrical phenomenon; in any case, he believed that because his mother was "a woman of genius and particularly excelling in the powers of intuition", there was a strong chance that he would witness paranormal phenomenon in association with her death.

In fact, Tesla did have a vision during the night that his mother died. He describes his brush with the "supernatural" in the excerpt below:

"But only once in the course of my existence have I had an experience which momentarily impressed me as supernatural. It was at the time of my mother's death. I had become completely exhausted by pain and long vigilance, and one night was carried to a building about two blocks from our home. As I lay helpless there, I thought that if my mother died while I was away from her bedside she would surely give me a sign[...] During the whole night every fiber in my brain was strained in expectancy, but nothing happened until early in the morning, when I fell in a sleep, or perhaps a swoon, and saw a cloud carrying angelic figures arvelous beauty, one of whom gazed upon me lovingly and gradually assumed the features of my mother. The appearance slowly floated across the room and vanished, and I was awakened by an indescribably sweet song of many voices. In that instant a certitude, which no words can express, came upon me that my mother had just died. And that was true. I was unable to understand the tremendous weight of the painful knowledge I received in advance, and wrote a letter to Sir William Crookes while still under the domination of these impressions and in poor bodily health."

Keep in mind that while Tesla had seen many strange things as a result of his unique talent for visualization, he had never been afflicted with hallucinations—images that appeared to him involuntarily were based on things he had seen in real life, and images of imagined things had to be laboriously created by his own effort. Tesla was a man of science, first and foremost, and as much as he might have wished to draw comfort from the idea that his mother had reached out to him from beyond the grave, he could not believe such a thing without putting it to test. After he began recovering from his physical exhaustion,

"I sought for a long time the external cause of this strange manifestation and, to my great relief, I succeeded after many months of fruitless effort. I had seen the painting of a celebrated artist, representing allegorically one of the seasons in the form of a cloud with a group of angels which seemed to actually float in the air, and this had struck me forcefully. It was exactly the same that appeared in my dream, with the exception of my mother's likeness. The music came from the choir in the church nearby at the early mass of Easter morning, explaining everything satisfactorily in conformity with scientific facts."

"This occurred long ago, and I have never had the faintest reason since to change my views on psychical and spiritual phenomena, for which there is absolutely no foundation. The belief in these is the natural outgrowth of intellectual development. Religious dogmas are no longer accepted in their orthodox meaning, but every individual clings to faith in a supreme power of some kind. We all must have an ideal to govern our conduct and insure contentment, but it is immaterial whether it be one of creed, art, science or anything else, so long as it fulfills the function of a dematerializing force. It is essential to the peaceful existence of humanity as a whole that one common conception should prevail."

In his adult life, Tesla had drawn away from the religious beliefs that had been imparted to him in his childhood. He continued to believe that religion, particularly Christianity and Buddhism, were excellent influences on society as a whole, as they gave people a moral system to guide their actions. But he did not believe in the afterlife or the supernatural as such. If there were ever a point in his life when he might have changed his mind about this, it must have been this one; but as his writing illustrates, even when facing the grief occasioned by his mother's death, scientific principles were the guide of all of his beliefs and convictions.

## Teleautomatics

No doubt Tesla's father, the Serbian Orthodox priest, would have taken some exception to his son's averring that superior intellectual development naturally precludes belief in religion. But in Tesla's case, this belief was more than just the natural disinclination of science to embrace spirituality; for him, it was all part of the theory of human automatism that Tesla had devised as a young man. When Tesla was trying to figure out how his powers of visualization worked as a child, he realized that everything that appeared to him in images were things he had seen in the course of his daily life; they appeared before his eyes, seemingly involuntarily, in obedience to various triggers, such as a word, a smell, or a sight that was associated with them on some level of his mind.

Tesla had, at first, no control over the visualization process; he was at the mercy of whatever environment triggers happened to influence him. This gave him cause to believe that everything that humans did, everything they thought, everything they felt or desired, occurred in response to some kind of stimulus. Therefore, Tesla concluded, humans were a kind of organic automaton—a robot, made of flesh—which was continually being programmed by influences which were sometimes perceptible, and sometimes mistaken for an idea originating in the human brain. As he puts it,

"We are automata entirely controlled by the forces of the medium being tossed about like corks on the surface of the water, but mistaking the resultant of the impulses from the outside for free will. The movements and other actions we perform are always life preservative and though seemingly quite independent from one another, we are connected by invisible links. So long as the organism is in perfect order it responds accurately to the agents that prompt it, but the moment that there is some derangement in any individual, his self-preservative power is impaired[...] A very sensitive and observant being, with his highly developed mechanism all intact, and acting with precision in obedience to the changing conditions of the environment, is endowed with a transcending mechanical sense, enabling him to evade perils too subtle to be directly perceived."


Tesla biographer Margaret Cheney writes that Tesla's theories on this subject seem unconvincing, as though the explanations he devised for the seemingly supernatural phenomenon he experienced did not entirely satisfy him. She points out that there were several instances in his life—according to the report of his extended family, his nephews and nieces—when Tesla appeared to manifest genuine precognitive intuition, such as when his sister Angelina was sick, and Tesla sent a message to his family expressing his worry that all was not well with her, prior to his being notified of her illness.

Tesla never seemed to think much of these episodes, but he could not fully explain them either, nor could he put his finger on what stimulus, precisely, the automata of his consciousness was responding to when he seemed to get wind of things before they happened. But perhaps it is not so surprising that Tesla, living an ocean away from his loved ones, would sometimes be seized with anxiety that misfortune had befallen them without his knowing, or that these anxieties would sometimes coincide with actual periods of illness or calamity. Life is uncertain; medicine in the last decades of the nineteenth century was still fairly primitive by today's standards, and Tesla's family lived in a relatively remote part of a comparatively underdeveloped country. The dangers of sickness beset them constantly, and Tesla's awareness of that fact probably needed no particular powers of prescience to explain it.

# Chapter Five: The Wizard of Fifth Avenue

## Native Son

After Duka Mandíc's death, Tesla became seriously ill from the cumulative effect of many months of overwork and exhaustion. He spent a few weeks in the mountains with his family, mourning the death of his mother and building up his strength after a year of dangerous exertion and taxing public lectures. When he began to recover his strength, he continued his lecture tour, this time giving speeches in Zagreb, the capital city of Croatia, and Belgrade, the capital city of Serbia. As an ethnic Serb who had grown up in Croatia, reinforcing this aspect of his identity was particularly important to him.

Tensions between Serbs and Croats would rise to a fever pitch within the next two decades, effectively sparking the first World War, but to Tesla, his Serbian identity was a source of inspiration to improve the lives of his countrymen and women by bringing the light of scientific advances to them. In Zagreb, his lecture was eagerly anticipated, but in Belgrade, he was met at the train station by hundreds of admirers, who lauded him as a hero. Standing on the railway platform, Tesla addressed the enthusiastic crowd:

"There is something within me that might be illusion as it is often case with young delighted people, but if I would be fortunate to achieve some of my ideals, it would be on the behalf of the whole of humanity. If those hopes would become fulfilled, the most exiting thought would be that it is a deed of a Serb. Long live Serbdom!"

Tesla won many awards throughout his life, and would have awards given in his name after his death, but the St. Slava Medal, awarded to him by the Serbian King Alexander I for special services to science, had special significance, and struck him with particular poignance after the king was assassinated eleven years later.

## The Chicago World's Fair of 1893

In January of 1893, shortly after Tesla returned to the United States and was again immersed in his experiments with electricity in his private laboratories, he received a telephone call from George Westinghouse, who had life-changing news: the Westinghouse company had beaten out Edison and General Electric, and been awarded the government contract for providing light and power at the Chicago World's Fair of 1893.

Also known as the World's Columbian Exposition, the fair was meant to celebrate the four hundredth anniversary of Columbus's discovery of America. In truth, it was in a more practical sense a spectacle that was intended to give people hope during a bleak depression. But the fair would make unprecedented use of electricity to light the enormous structures, some modeled on famous European landmarks, others like nothing the world had ever seen before—including the world's first Ferris wheel.

The fact that Westinghouse had been awarded the contract signaled a decisive shift in the so-called "War of the Currents". Not only would the entire exposition be run on alternating current power, but the president of the United States, Grover Cleveland, had agreed to perform the formal flip of the switch that would mark the fair's official opening. This was to be a significant moment in the history of electricity—for years, Edison and his propaganda team had been warning the public in the most dramatic terms that to touch a switch connected to alternating current power was to risk death. Electric lighting had been installed in the White House, but the President himself was not permitted to operate the light switch, as the risk of a short circuit and death by electrocution was deemed too great. The fact that Cleveland was willing to operate a much, much more powerful switch before the eyes of the entire city of Chicago (not to mention a number of important guests, such as European royalty) signaled that alternating current power was about to enjoy a new age of legitimacy.

It was with considerable reluctance that Tesla had pulled his head out of his personal experiments with electricity to help George Westinghouse spread the good news about alternating current at the World's Fair, but once he had turned his mind to the task, the results did not disappoint. Tesla did not much resemble our contemporary stereotype of the withdrawn, socially awkward scientific genius who has little warmth for anything not grown in a petri dish. Tall and lean at six foot six and one hundred and forty four pounds, he was both handsome and striking; his dark hair and mustache were neatly groomed, and his dress extremely neat, even on an ordinary day. When presenting his scientific exhibitions, Tesla wore a white tie and tails; he resembled a magician, with something much more exciting than a rabbit to pull out of his hat.

Tesla's exhibits at the Chicago World's Fair of 1893 included items that became commonplace in the latter half of the twentieth century, such as phosphorescent tube lighting, the precursor to fluorescent lights. One can imagine how striking it must have been for the very first people ever to lay eyes on a delicately hand blown length of tube lighting, molded to spell out words and phrases, such as "Lights" and "Welcome, Electricians". Of course, the majority of the people who filed through the Electricity

Building—where Thomas Edison was also holding court, the central feature of his display being his sixty foot high "Tower of Light"—were not electricians, and had absolutely no grasp of the scientific or engineering principles behind the fantastic displays they were witnessing. But as Tesla channeled currents through his own body, causing his clothes and hair to emit sparks, and made eggs spin in place, his audiences were captivated nonetheless.

Those who witnessed Tesla's display rooms in the Electricity Building may or may not have been aware that it was Tesla's polyphase system that was powering the entire fair, including the so-called White City (a cluster of temporary buildings made of plaster and other degradable fibers, covered in white stucco so as to reflect the street lighting), and the replica Venetian canals. This incredible feat would open the door to other government contracts and highly visible projects for Tesla, Westinghouse, and alternating current power in later years.

## Back to New York: Tesla in Society

A side effect of Tesla's now immense fame was that the world of New York high society was open to him. Late nineteenth century society was in a period known as the "Gilded Age"—the age of the new American millionaire, industrialists and robber barons who snapped up natural resources and emerging technologies and commodities to create monopolies, amass incredible wealth, and shut down all competition. These were the Vanderbilts, the Astors, the Carnegies, and the Morgans, among others: at the head of each family, there was a scheming captain of industry who had built his fortune on the backs of hundreds of thousands of men in work-gangs, laboring twelves hours a day for a few cents of pay. And at their sides were the colorful society hostesses who set fashions, issued invitations, and created the social tableaus that provided the backdrop for in-fighting, deal-making, and match-making among New York's most powerful families.

It was not a world to which Tesla was born, but it was one with which he had to become familiar. He had freely given up his one opportunity to become a multi-millionaire himself, choosing instead to preserve George Westinghouse's company; and while he did not yet lack for money to keep himself in decent suits and pay the rent on his laboratories, he required rich investors if he was to keep making inventions indefinitely. And the rich men of Wall Street were very interested in him. He was the famous genius whose brain had produced the induction engine, and men like that existed to make money for men like Vanderbilt and Morgan. It just so happened that in the mid 1890's, Tesla was living at the Waldorf-Astoria, a hotel next door to the

New York Stock Exchange, where the money men retired after the close of business to have a few drinks and rehash the business of the day. Tesla, famous, handsome, eccentric, but charming, was soon an accepted member of their ranks. They were not very interesting people to talk to, but they were necessary to advancing his career.

Tesla was not at all unappreciated in his time. For a few decades after his death, his name was comparatively forgotten in comparison to Thomas Edison, who in addition to being an inventor also took the precaution of securing his legacy through business, and thus laid eternal claim to the affection of American historians. But during his lifetime, Tesla was spoken of everywhere, written of in newspapers, and befriended by celebrities, such as American novelist Mark Twain. Nowadays, most scientists are either academics, employed by universities, or employed by government agencies or industrial conglomerates. Tesla was merely a man with a laboratory, who liked to show people what he could do with electricity, and this has contributed to the decline of his reputation over the decades. But in the nineteenth century he was widely adored; the natural repercussion of this was that he was also intensely hated, mostly by people who had never met him and therefore assumed that the feats attributed to him by the newspapers could only be elaborate hoaxes.

Tesla could not afford to continue living at the Waldorf indefinitely. Eventually he moved into the Gerlach Hotel, a decidedly no-frills establishment that scarcely suited his tastes. But he continued to have access to the social world of the super-rich in New York through his close friends, Robert and Katharine Johnson. Robert Johnson was a writer and a diplomat, and though not particularly wealthy themselves, the couple had an extraordinary ability to attract the friendship of those above them; to Tesla, they extended the advantage of their social talents.

Throughout Tesla's life, he was the subject of serious gossip regarding his personal relationships, or lack thereof. The very rich people with whom he associated were also very bored, and match-making was one of their favorite sports. And in addition to this, marriage was not seen as optional, even for men, in the 1890's. Tesla was immensely popular, and terribly handsome—why did he not marry? It was the subject of speculation in newspaper columns, who regarded his bachelor condition as "unnatural". After the trial of Oscar Wilde in 1895, a wave of paranoia spread through English (and by automatic extension, American) society regarding male homosexuality. Even men who were not gay, particularly the unmarried ones, were wary of falling under suspicion of having committed "indecent acts" if they seemed too familiar with their male friends, while gay men, though they had always lived in secret, faced increased danger of harsh sentencing and complete social disgrace if they were caught in an illegal relationship. Even Tesla's friends and admirers speculated as to why he was not married, while his detractors, including those from the Edison camp and those motivated by mere professional jealousy, had no qualms about insinuating that he was homosexual.

The truth is that there is very little evidence indicating what Tesla's sexual orientation might have been. All that is known is that he never married, never courted a woman publicly, and only engaged in light flirtations as a social amusement. One or two incidental factors have roused the curiosity of historians as to what kind of relationships he might have pursued out of the public eye. He once seemed to develop a close friendship with a young man whom he met at the home of Robert and Katharine Johnson, a handsome young naval officer who seemed to embody Tesla's romantic ideal of what a man should be. Their friendship was fuel for the rumors that Tesla was gay, but there is no recorded proof of their having had such a relationship, or of Tesla nurturing romantic feelings for him. On another occasion, Tesla intimated to a friend that he maintained a small apartment some blocks away from his primary residence where met with "special friends" in private. There is no way of telling whether those meetings were romantic or sexual in nature, or, if they were, whether the "special friends" Tesla referred to were men or women.

If Tesla were gay, that might not necessarily have stopped him from marrying, any more than it did Oscar Wilde, who had a wife and two sons when he was convicted and sentenced to two years' hard labor for gross indecency. But for Tesla, a lack of attraction to women, if that was indeed the problem, was not the only obstacle to marriage. His obsessive-compulsive behaviors, germphobic tendencies, and violent, irrational dislike of random things, would have constituted a far more impenetrable barrier to intimacy—with persons of any gender, one would think. The sight of a pair of earrings on a woman bothered him so intensely that he could barely hold a conversation with her; other kinds of jewelry bothered him as well. And he wrote in his autobiography that he would only touch the hair of another human being, perhaps, at the point of a pistol. All of this, taken together with Tesla's preference for losing himself in his electrical experiments, working twenty hour days, and sleeping (supposedly) no more than two hours a night, make it seem quite understandable that he never married, and would make it unsurprising if he never felt comfortable having a sexually intimate relationship with any woman or man.

Finally, on the subject of marriage, Tesla had this to say to an interviewer who asked him if "persons of artistic temperament", such as Tesla, ought to marry:

"For an artist, yes; for a musician, yes; for a writer, yes; but for an inventor, no. The first three must gain inspiration from a woman's influence and be led by their love to finer achievement, but an inventor has so intense a nature with so much in it of wild, passionate quality, that in giving himself to a woman he might love, he would give everything, and so take everything from his chosen field. I do not think you can name many great inventions that have been made by married men."

Thomas Edison, of course, had been married twice, but it is a matter of personal interpretation whether Tesla intended a personal slight.

## The Niagara Falls Commission

For years, a group of scientists and electrical engineers had debated how to harness the potential power generated by the Niagara Falls to produce electricity. Called the Niagara Falls Commission, the committee had offered a prize of three thousand dollars around the time Tesla first came to the United States to be awarded to the engineer who could present a viable method for harnessing the power of the falls. George Westinghouse, declining to enter the competition, had remarked that the Commission was attempting to purchase a hundred thousand dollars' worth of information for three thousand dollars, and that the Commission could call him "when they were ready to talk business". In October of 1893, fresh from the triumphs of lighting the Chicago World's Fair, Westinghouse received that call—despite all of Edison's efforts, alternating current power had become the acknowledged professional standard of the day, and Westinghouse was still its primary distributor.

Word of Westinghouse's victory could only come as a piece of marvelous good news to Tesla, who had been contemplating the problem of harnessing the Niagara Falls since he was a small child. As he writes in his autobiography,

"In the schoolroom there were a few mechanical models which interested me and turned my attention to water turbines. I constructed many of these and found great pleasure in operating them. How extraordinary was my life an incident may illustrate. My uncle had no use for this kind of pastime and more than once rebuked me. I was fascinated by a description of Niagara Falls I had perused, and pictured in my imagination a big wheel run by the Falls. I told my uncle that I would go to America and carry out this scheme. Thirty years later I saw my ideas carried out at Niagara and marveled at the unfathomable mystery of the mind."

The harnessing of the falls was accomplished by the building of a powerhouse containing generating units—seven in total, completed by Westinghouse, and generating fifty thousand horsepower of electricity all together. In time, Edison's General Electric company came to share the contract and build a second powerhouse, containing eleven generators. But Edison's victory can only have been bittersweet; General Electric had been obliged to buy a license to use Tesla's patented polyphase system, and thus the generators that Edison's company built for the Niagara Falls Commission was run on alternating current.

Tesla won a number of awards and honors based on the technology used in the Niagara Falls

Commission project. There was a limit to how much he was able to enjoy them, however. An upsurge in anti-Tesla sentiment followed the wave of publicity the project generated. Tesla's patents were constantly being challenged in court by inventors who claimed that they had invented virtually identical machines and components before him. Others simply used Tesla's designs without buying the license, and had to be sued in turn by Westinghouse, to whom Tesla had sold his patents, to protect the proprietary technology. Every single suit brought against Tesla's patents was upheld by the courts; two of them made it as far as the United States Supreme Court, and were upheld there as well. But the backlash of negative publicity generated by the legal conflict made the public unsure what the real story was, and the opinion of the scientific community became divided. Disheartened, Tesla returned to his laboratories and buried himself in his experiments once again. He was always happiest there, and only a desire to better the world by promoting acceptance of his inventions had persuaded him to leave in the first place.

**End of An Era**

When Tesla returned to his workshop experiments after the Niagara Falls Commission project in 1893, he was able to spend some months in happy scientific absorption. He was again drawn into taxing work habits, though not perhaps so utterly to the ruin of his health as in the year before. He had everything he needed for happiness: enough money to keep him in modest comfort, a laboratory full of experiments, and the company of friends, when he desired it. (Actually, his friends requested his company far more often than he deigned to give it.)

Tragically, this happy and contented period of his life came to an end after about two years. In

March of 1895, a fire destroyed the building on Fifth Avenue that contained Tesla's laboratory and all of his equipment and research. The loss was total; nothing could be salvaged, nothing was insured, and even if it had been, no amount of money could have compensated Tesla for the loss the experiments that were still in-progress. Tesla was understandably devastated, and the newspapers, most of which still venerated Tesla as a heroic genius, expressed the loss felt by the scientific community at the thought of all the marvelous inventions that were forever lost to the fire:

"The destruction of Nikola Tesla's workshop, with its wonderful contents, is something more than a private calamity. It is a misfortune to the whole world. It is not in any degree an exaggeration to say that the men living at this time who are more important to the human race than this young gentleman can be counted on the fingers of one hand; perhaps on the thumb of one hand."

Tesla biographer Margaret Cheney gives us a more detailed estimate of the extent of what was lost to science and the world when Tesla's experiments were destroyed:

"Only [Tesla's] closest assistants knew the dazzling scope of his advanced researches in radio, wireless transmission of energy, and guided vehicles, or that he was achieving effects with what the world would soon know as X rays, and also nearing a breakthrough in the potentially lucrative industrial discovery of a means of producing liquid oxygen. It may have been the latter volatile substance that caused the blaze—apparently started from a gas jet on the first floor near oil-soaked rags—to explode so rapidly through the building."

Adding to Tesla's many griefs was the prosaic problem of money. Having given up his rights to any royalties from his patents when he sold them to George Westinghouse, and having spent all of the money he accrued from salaries and investors on his equipment, he was essentially without capital. With the exception of a few remaining German patents that were still in his possession, Tesla was now completely wiped out, without any clear means of rebuilding his life or his work.

# Chapter Six: Houston Street and Beyond

"Immediately after the destruction of my laboratory by fire, the first thing I did was to design this oscillator (shown in fig. 27). I was still recognizing the absolute necessity of producing isochronous oscillations, and I could not get it with the alternator, so I constructed this machine. That was all a very expensive piece of work. It comprised four engines. Those four engines were put in pairs and there was an isochronous controller in the center, and furthermore, that controller was so arranged that I could set two pairs of engines to any phase or produce any beat I desired. Usually I operated quarter phase; this is, I generated currents of 90° displacement.

"By the way, now, for a first time you see my apparatus on Houston Street, which I used for obtaining oscillations, dampened and undampened as well. But it was necessary to state that while others, who had been using my apparatus, but without my experience, have produced with it dampened oscillations, my oscillations were almost invariably continuous, or undamped, because my circuits were so designed that they have a very small dampening factor. Even if I operated with very low frequencies, I always obtained continuous, or undampened, waves for the reason that I designed my circuits as non-radiative circuits."

Nikola Tesla, from an interview for a legal suit over his patents, 1916

After the destruction of his Fifth Avenue labs, Tesla was left to rely on the one source of wealth he could never be parted from: his extraordinary powers of memory and visualization, which had preserved most of his research at exactly the point where it had left off, with no more loss of detail than if he had managed to preserve his notes, plans, and drafting papers. And there was another perennial resource available to him: the vested interest of wealthy men who recognized Tesla's inventive genius, and knew that if they backed him, their investment was likely to pay extraordinary dividends.

Tesla's first priority was to build a new laboratory and replace all of the equipment that had not been specially invented by himself and thus could be fabricated by other manufacturers. To his relief, a sponsor soon presented himself in the form of Edward Dean Adams, one of the financiers behind the Niagara Falls Commission project, who also had business ties to the ludicrously wealthy American industrialist James Pierpont Morgan. With a sizable cash sum from Adams in hand, and equipment from George Westinghouse, Tesla was able to establish a new laboratory for himself on Houston Street in New York.

He was not, however, interested in the other offer made to him by Adams, which was to form a new company that would have the direct backing of Morgan. Tesla believed, not without good reason, that this would lead to Morgan interfering with his research, dictating the projects he worked on, and possibly even using his inventions for purposes he did not approve of. From a financial point of view, a deal with Morgan would have meant long term financial security, the likes of which Tesla was never to have again. But from Tesla's perspective, refusing to get into bed with Morgan was a matter of principle, and considering Morgan's character and reputation, it is difficult to believe that he was wrong to do so.

**X-Rays**

In 1895, a monumental discovery rocked the scientific world: Wilhelm Röntgen, a German engineer and physicist, discovered a new wavelength of electromagnetic radiation that produced photographs which revealed the skeleton within the human body. These were called Röntgen rays, but soon became known simply as X-rays.

The thing about Tesla's having invented so many electrical machines simply for the joy of seeing what they could do was that he sometimes produced groundbreaking effects by accident without quite knowing what he had discovered. One of these accidental effects was a photograph which Tesla had attempted to take of his friend, the famous author Mark Twain, in 1894. The picture was not made with an ordinary camera, but with the light produced by a Geissler tube, and the photograph it produced was not an image of Twain at all, but of the inside of the camera itself.

After Röntgen's discovery made headline news, Tesla, innocently delighted, sent a copy of this photograph to Röntgen; not in an attempt to lay a prior claim to his discovery (though, arguably, the photo was proof that X-rays were another discovery that Tesla had anticipated), but simply to share his interesting findings. Tesla continued to experiment with X-ray photography for some time after Röntgen's announcement. In its infancy, X-ray technology required very long exposure times in order to produce clear images, not unlike early visible light photography some fifty years prior, which required the subjects to pose for ten or twenty minutes at a time to produce a good quality image. In the course of Tesla's purely exploratory X-ray research, he made a remarkable claim:

"I am producing strong shadows at distances of 40 feet. I repeat, 40 feet and even more. Nor is this all. So strong are the actions on the film that provisions must be made to guard the plates in my photographic department, located on the floor above, a distance of fully 60 feet, from being spoiled by long exposure to the stray rays. Though during my investigations I have performed many experiments which seemed extraordinary, I am deeply astonished observing these unexpected manifestations, and still more so, as even now I see before me the possibility, not to say certitude, of augmenting the effects with my apparatus at least tenfold!"


"These effects upon the sensitive plate at so great a distance I attribute to the employment of a bulb with a single terminal, which permits the use of practically any desired potential and the attainment of extraordinary speeds of the projected particles. With such a bulb it is also evident that the action upon a fluorescent screen is proportionately greater than when the usual kind of tube is employed, and I have already observed enough to feel sure that great developments are to be looked for in this direction".

Tesla biographer Margaret Cheney points out that if the claims Tesla makes above are true, "he would have been using equipment far more advanced than anything we now believe existed at that time."

Rather like Marie Curie, who died of aplastic anemia after exposing herself to fatal levels of radiation when she was pioneering the world's very first experiments with radium, Tesla and the other X-ray experimentalists were taking serious risks with their own health. But of course, the dangers of X-rays would not be fully understood until years later, partly thanks to Tesla himself. When he first began his research, Tesla was convinced that X-rays were harmless to the human body—in fact, he believed they stimulated brain activity, and as a consequence he radiated his own head on a regular basis. He also radiated other body parts—eyes, hands, skin—and gradually began to take notice of burns, discolorations, and other damaging effects that the X-rays produced. As a result of his observations, Tesla began to give lectures on the necessity of taking safety precautions when working with X-rays, and it is as a result of this that lead aprons began to be used to shield experimenters from the effects of repeated X-ray exposure. Unfortunately, these precautions did not become common until after some X-ray researchers had already suffered disastrous consequences—Thomas Edison damaged his eyesight permanently by working with X-rays, and one of his assistants, Clarence Dally, developed a skin cancer that spread over time and eventually proved fatal. Edison ceased working with X-rays entirely because of this.

## Mars and Earthquakes

Tesla believed it would soon be within humanity's power to communicate through wireless transmitters (amplified by the earth itself) with the planet Mars—where, he was certain, a race of intelligent beings lived. (He referred to the likelihood of life on Mars as a "statistical certainty"; this was long before much was yet understood about the size of the galaxy, or the number of galaxies in the universe.) His fellow scientists considered his idea ridiculous, but Tesla stood by the theory.


On another occasion, Tesla was testing the effects of an electromechanical oscillator by connecting it to an iron pole that ran through his Houston Street laboratory, down into the basement of his building. Tesla kept increasing the frequency, noting how various items around him responded, rather like a bean jumping in a hot skillet. To Tesla's perception, the effects of the oscillator were fairly mild—he was unaware that the resonance was traveling down the iron pole and spreading throughout his New York neighborhood, causing buildings to shake and windows to break. As New York is not precisely known for its earthquakes, the police promptly got in touch with Tesla at his lab, which was already known to be the likely source of any strange thing that was happening in the neighborhood. The officers arrived just as Tesla noticed what was happening and smashed the oscillators to pieces in order to stop the effects as quickly as possible.

Tesla was conscious of the potentially devastating side effects of wielding this much power. He claimed in a newspaper interview that he once clamped a vibrating resonance device to a steel beam at a construction site, and nearly succeeded in making the entire building collapse. Tesla claimed that he could use a similar device to destroy the Brooklyn Bridge—or do something even more catastrophic. A lengthy newspaper article written by Allan L. Benson in 1915, when

Tesla was fifty nine, goes into more detail about the supposedly immense destructive powers of his oscillators:

"Tesla says that he can split the earth in the same way—split it as a boy would split an apple—and forever end the career of man. This seems like quite a large order—but—see what he says about it.

"The vibrations of the earth,' said he, 'have a periodicity of approximately one hour and forty-nine minutes. That is to say, if I strike the earth this instant, a wave of contraction goes through it that will come back in one hour and forty-nine minutes in the form of expansion. As a matter of fact, the earth, like everything else, is in a constant state of vibration. It is constantly contracting and expanding. Now, suppose that at the precise moment when it begins to contract, I explode a ton of dynamite. That accelerates the contraction and, in one hour and forty-nine minutes, there comes an equally accelerated wave of expansion. When the wave of expansion ebbs, suppose I explode another ton of dynamite, thus further increasing the wave of contraction. And, suppose this performance be repeated, time after time. Is there any doubt as to what would happen? There is no doubt in my mind. The earth would be split in two. For the first time in man's history, he has the knowledge with which he may interfere with cosmic processes.'"

"As usual," observes Margaret Cheney, "Tesla's comments to the press smack of exhibitionism." Reading his interviews, it certainly does seem that Tesla relished making dramatic statements; but if one reads further, Tesla eventually admits that while the theory "could not fail to work", it would be impossible to build a machine that aligned with the earth's resonances so precisely.

## Robots and the Spanish American War

In 1898, Tesla revealed to a reporter a new invention which he declared would be a free gift, given to the world for its betterment—an invention which he could not describe in too much detail, lest other inventors try to patent it, and thus prevent it being used to help the people of the earth, but which he was willing to let the reporter have a glimpse of. This invention turned out to be the solar panel: the invention with which Tesla intended to "harness the rays of the sun" and revolutionize how energy was generated.

Tesla's goal was, as always, noble. He wanted to do away with the need to mine for fossils fuels, a process which devastated the earth and imperiled the lives of the impoverished workers employed as miners. But Tesla was having the same problem with his solar panels that he was having with most of his inventions. His true love was chasing down an idea as it spun off in unexpected directions; he had a harder time shaping that idea into a form that could be made to serve a commercial purpose, and therefore yield patents that would make him any money.

The fact that making money was never the first goal on Tesla's mind was put into evidence in an even more dramatic fashion, through a demonstration that he gave when the Spanish-American War broke out in 1898. Villifying Spain over its rule of Cuba was an enormously popular cause amongst the elite of New York society; in fact, William Randolph Hearst, the newspaper baron, was instrumental in provoking the United States Congress to declare the war, by printing false stories of Spanish cruelty towards heroic Cuban rebels, and inciting the American public to demand that the United States flock to the rebels' aid. Hearst's goal was to boost the flagging sale of newspapers, but he touched off a romantic firestorm amongst people like Theodore Roosevelt, who recruited a band of upper class New York men to form the "Rough Riders" volunteer brigade, which joined the fight for Cuban independence without anyone, especially the Cubans, asking them to do so. Essentially all of Tesla's friends and colleagues amongst his Manhattan social circle were invested in the war, and Tesla was not far behind them.

At an exhibition in Madison Square Garden, Tesla revealed to the world his prototype for the very first remote controlled boat—a submersible, which could, via radio waves, be deployed without a crew, and be made to sail, sink underwater, transmit information, and fire torpedoes, with endangering a single human life. (At least, on the side of the nation that controlled the boat.) Tesla was an exhibitionist by nature, but he was unusually secretive about the plans for his robotic boat, even when he filed the patent. He told reporters that he was keeping the full details of the boat plans, and his latest advancements in wireless transmissions, a secret, because he intended to offer them to the United States government to assist them in the war effort.

And in fact, Tesla did just this: he explained the robotic boats' full functionality to members of the American navy, along with his projected estimate that each boat could be completed for around fifty thousand dollars. However, his plans were rejected—mostly on the basis that they sounded too good to be true. The top military leadership in the United States were not, as a body, inclined to adapt cutting edge scientific breakthroughs in their plans for war. This would change in a big way during the second World War. Military leaders in later decades, including senior naval personnel who were involved in the planning and operation of World War II, give credit to Tesla for making the first strides into the remote-operated military weapons that are a standard part of the arsenal of war today.

Tesla's vision of a future made better by robotics extended to remote controlled aircraft and remote controlled automobiles. "By installing proper plants," he said, "it will be practicable to project a missile of this kind into the air and drop it almost on the very spot designated, which may be thousands of miles away. But we are not going to stop at this. Teleautomata will be ultimately produced, capable of acting as if possessed of their own intelligence, and their advent will create a revolution." In this present age of drones and the first self-driving cars, this sounds like more confirmation of Tesla's powers of foresight.

# Chapter Seven: Colorado

## The Colorado Springs Laboratory

Just as Tesla abandoned any number of other projects without bringing them to the point of commercial readiness, he turned his back on teleautomatry and robotics; not because he was insufficiently interested, but because, given the reaction of the U.S. military and the non-reaction of the general public, he could tell that the world was insufficiently prepared to appreciate the scope of what he had created. He moved on to other experiments and other lines of research—and it soon became apparent that he was going to need a new lab.

More wary than ever of the risk of a fire breaking out, Tesla wanted a research space built to his specifications, with a high ceiling to accommodate the sparks that jumped out from his devices and climbed along the ceiling. One presumes that he was also mindful of the fact that he had triggered a small earthquake in the heart of New York City; he not only needed greater amounts of space inside his lab, he needed a healthy amount of space outside and surrounding his lab, to protect innocent bystanders.

Tesla's main research goal in the spring of 1899 was the worldwide wireless system, his longstanding project dedicated to finding a means of transmitting electrical energy without wires. His hope was to build an enormous transmitter which would be able to send a signal all the way to the coast of Cornwall, in England. By building the transmitter tall enough, his theory went, it might be easier to transmit energy, because it would be passing through the upper levels of the atmosphere, where the air was much thinner.

In searching for his new research space, Tesla applied to his patent lawyer, Leonard Curtis, who made swiftly made arrangements to help him. Curtis had interests in the Colorado Springs Electric Company; so to Colorado Springs, Curtis invited Tesla to come, by a telegram that read: "All things arranged, land will be free. You will live at the Alta Vista Hotel. I have interests in the City Power Plant so electricity is free to you." This was precisely what Tesla wanted to hear. Since leaving Edison's employ, he had always depended on powerful or well-connected friends to clear away the obstacles in the path of pursuing his genius, and Curtis, who was also Tesla's most dedicated defender in the legal realm, was performing his role admirably. Other powerful friends were also combining forces to enable Tesla's research: New York billionaires like Colonel John Jacob Astor were putting their financial backing behind the new laboratory.

The land given to Tesla for the construction of his laboratory was about a mile away from the city of Colorado Springs, a site which had primarily been used for grazing for dairy cows, close to a state school for deaf and blind students. Tesla was extremely pleased by the location. As he described it,

"The conditions in the pure air of the Colorado Mountains proved extremely favorable for my experiments, and the results were most gratifying to me. I found that I could not only accomplish more work, physically and mentally, than I could in New York, but that electrical effects and changes were more readily and distinctly perceived."

He was transported back and forth from his hotel to the laboratory site by an open horse carriage, still the primary means of transportation in that part of the country at the turn of the century.

Cheney describes the great work Tesla was arranging this immense research space to develop, and the impact it was still having on the world at the time she was writing in 1981:

"This transmitter, which he developed in Colorado, he would later claim as his greatest invention. Indeed, it is the Tesla invention that continues to fascinate many of his modern followers the most. Whenever and wherever in recent years phenomena have been detected, resulting from powerful radio signals pulsed at very low frequencies, journalists speak knowingly of the Tesla effect. The Russians, it has been claimed, are using a giant Tesla magnifying transmitter to modify the world's weather, creating extremes of ice and drought. It is said to cause periodic disruption of radio communication in Canada and the United States with attendant brain-wave interference and vague symptoms of physical distress, not to mention sonic booms and almost anything else not otherwise explicable."

Tesla described his transmitter as a "resonant transformer" that is "suitable for any frequency, from a few to many thousands of cycles per second, and can be used in the production of currents of tremendous volume and moderate pressure, or of smaller amperage and immense electro-motive force. The maximum electric tension is merely dependent on the curvature of the surfaces on which the charged elements are situated and the area of the latter."

Tesla ultimately intended for his transmitter to be adapted for commercial purposes, but not until after years of experimentation had been completed. And either because this project was relatively much more important to him than any project he had worked on before, or simply because he was starting to crest middle age and his memory was no longer as efficient as it had been, Tesla kept copious notes and took many photographs of his research. Always before, he claimed, he had merely visualized his inventions, made adjustments to his internal schematics, and built the final working model when it was already perfectly formed in his mind. But since the notes he left behind from his Colorado Springs laboratory are only partially intelligible to modern scholars, it seems certain that he was still doing the majority of his calculations in his head.

It is just as well that Tesla did not attempt his transmitter research in the middle of New York, because he routinely electrified the Colorado desert for miles around his research space. Every life form in the vicinity, from insects to horses, reacted to the atmospheric disturbances Tesla produced. There were flashing lights, sounds and colors, sparks dancing between grains of sand—if Tesla had not already gained such a widespread reputation for conducting earth-shaking experimentation, it is probable that people in the nearby towns would have suspected that Armageddon had arrived. As it was, his neighbors would simply come out of their homes to watch the atmospheric disturbances he created—yet another one of Tesla's spellbinding, theatrical, magical demonstrations.

**Stationary Waves**

In July of 1899, during a particularly violent thunderstorm that cracked open the Colorado skies above his research station, Tesla had a revelation which, he believed, was among the most important for the future of the human race he had ever uncovered in his career. He described the ebb and flow of electrical activity during the storm, rolling electrical pulses that were stronger, then weaker. He found himself reaching a conclusion as he tracked the pattern of the pulses: they constituted something called stationary waves, and their implications were far-reaching. Tesla wrote,

"Impossible as it seemed, this planet, despite its vast extent, behaved like a conductor of limited dimensions. The tremendous significance of this fact in the transmission of energy by my systems had already become quite clear to me.

"Not only was it practicable to send telegraphic messages to any distance without wires, as I recognized long ago, but also to impress upon the entire globe the faint modulations of the human voice, far more still, to transmit power, in unlimited amounts to any terrestrial distance and almost without loss."

Some of the experiments Tesla carried out in the Colorado Spring laboratory were immensely dangerous, and stood at genuine risk of killing Tesla or his assistants or burning the research station down—again. In his most extreme experiment, his artificial lightning shot 135 feet high, before the coil went dead. Thinking that his assistant had turned off the power switch, Tesla ordered him to switch it back on, only to be told that the generator that supplied their electricity was dead. When Tesla phoned the power company to demand that their power be restored, he was curtly informed that his experiment had blown out the company's dynamo, and now it was on fire. There were no electric lights in Colorado Springs that night.

Much of what Tesla invented at his Colorado Springs laboratory never made it past his own notes, and modern scholars are still attempting to unravel the details of all that he accomplished.

## Sounds from Outer Space

One late night at the Colorado Springs laboratory, Tesla was alone and at work when he began to detect a very faint signal coming through his transceiver—a signal that seemed to be coming from outer space. Tesla had longed believed in the "mathematical certainty" of the existence of extraterrestrial life, and was convinced that Mars must be inhabited and attempting to communicating with Earth. Where this signal was coming from, be it Mars or elsewhere, he could not possible tell, but he wished to make some sort of reply, if possible. He told the story to journalist Julian Hawthorne in 1900:

> "Apart from love and religion there happened the other day to Mr. Tesla the most momentous experience that has ever visited a

human being on this earth. As he sat beside his instrument on the hillside in Colorado, in the deep silence of that austere, inspiring region, where you plant your feel in gold and your head brushes the constellations — as he sat there one evening, alone, his attention, exquisitively alive at that juncture, was arrested by a faint sound from the receiver — three fairy taps, one after the other, at a fixed interval. What man who has ever lived on this earth would not envy Tesla that moment! Never before since the globe first swung into form had that sound been heard. Those three soft impulses, reflected from the sensitive disc of the receiver, had not proceeded from any earthly source. The force which propelled them, the measure which regarded them, the significance they were meant to convey, had their origin in no mind native to this planet.

"They were sent, those marvelous signals, by a human being living and thinking so far away from us, both in space and in condition, that we can only accept him as a fact, not comprehend him as a phenomenon. Traveling with the speed of light, they must have been dispatched but a few moments before Tesla, in Colorado, received them. But they came from some Tesla on the planet Mars!"

Tesla did not widely publicize this experience at first, knowing that there were dedicated anti-Teslans in the world who would jump to point to this unusual experience as proof that he was going mad (or possibly that he had been mad all along). But scientists and amateur astronomers searching for extraterrestrial life these days look back on Tesla's experience with the signals from outer space as the first attempts of humanity to reach out to life beyond our planet.

# Chapter Eight: Radio

**Tesla vs. Marconi**

The rivalry between Tesla and Marconi has taken on legendary proportions. Even now, when both inventors have been dead for some time and the medium of radio itself is giving place to various forms of digital communications, the question of whether Nikola Tesla or Guglielmo Marconi was the true original inventor of radio can still get the blood of both men's advocates pumping. (Amusingly, the first page of Google search results for the phrase "Tesla vs Marconi" include websites titled "Nikola Tesla: The Guy Who DIDN'T 'Invent Radio'", and "Tesla Invented Radio, Not Marconi!")

As this book has mentioned before, Tesla, though not the perfect model of the mad scientist in every way, did fit the image of the genius who is too distracted by his discoveries and experiments to pay as much attention as he should to practical matters, like his finances and his patents. Fortunately, he had the good sense to begin fielding responsibility for his patents to lawyers and assistants. The most intense patent battle of his life (in fact, it was not settled until a few months after his death) was with Guglielmo Marconi, over the radio transmitters.

As early as 1892, Tesla had realized the theoretical possibility of transmitting radio signals over increasingly long distances. In 1898, he attempted to demonstrate the practical applications of radio transmission to members of the United States military by showing them the first radio controlled boats. He had foreseen, from the beginning, the possibilities of transmission distances that crossed oceans; but as with so many of his other experiments, Tesla was more interested in theory, which generated an increasing number of suggestive possibilities for new applications, than in hammering out the nuts and bolts of his ideas until they could be developed and sold commercially. Marconi, however, was more or less the opposite; he was a little more like Edison, as much of a businessman as an inventor, though with more Tesla-style inspiration on his side than Edison had on his. He was enormously wealthy, as a result of the manufacturing enterprise he built upon around his inventions (for which he, unlike Tesla, kept control of the patents and the royalty payments.)

What Marconi did that Tesla did not do was concentrate steadily on extending the range of the broadcast signal, first over a distance of one mile in 1901, then over a distance of a hundred miles. By 1902, he was sending radio transmissions across the Atlantic ocean, from England to Newfoundland.

The most balanced evaluation of the Tesla versus Marconi dispute appears to place the verdict somewhere between the two inventors' claims. Toward the end of his life, Tesla sued Marconi over the patents on some of the equipment he used to develop and increase the range of his signals, but the essential difference between the two inventors was that Tesla was a theoretical scientists first and foremost, while Marconi concentrated on engineering something practical and commercially viable based on the essential principles that other scientists had come up with first. The fact of the matter is, Tesla could easily have "invented radio"—that is, made a device that could transmit signals and messages easily over long distances—several years before Marconi, if that had been where he was focusing his concentration. And he attempted to, when he built Wardenclyffe Park, which we will discuss in an upcoming section. But goals were never singular; he was interested in much more than strengthening broadcast signals, and as a result,

Marconi produced the first set of tangible results.

In 1917, John Stone presented the Edison Medal to Nikola Tesla on behalf of the American Institute of Electrical Engineers, and the following excerpt from his speech gives some indication of how Tesla's contributions to the field of radio were regarded in the early part of the century:

> "I misunderstood Tesla. I think we all misunderstood Tesla. We thought he was a dreamer and visionary. He did dream and his dreams came true, he did have visions but they were of a real future, not an imaginary one. Tesla was the first man to lift his eyes high enough to see that the rarified stratum of atmosphere above our earth was destined to play an important role in the radio telegraphy of the future, a fact which had to obtrude itself on the

attention of most of us before we saw it. But Tesla also perceived what many of us did not in those days, namely, the currents which flowed way from the base of the antenna over the surface of the earth and in the earth itself."


"Tesla, with his almost preternatural insight into alternating current phenomenon that had enabled him some years before to revolutionize the art of electric power transmission through the invention of the rotary field motor, knew how to make resonance serve, not merely the role of a microscope to make visible the electric oscillations, as Hertz had done, but he made it serve the role of a stereopticon to render spectacular to large audiences the phenomena of electric oscillations and high frequency currents....He did more to excite interest and create an intelligent understanding of these phenomena in the years 1891–1893 than anyone else, and the more we learn about high frequency phenomena, resonance, and radiation today, the nearer we find ourselves approaching what we at one time were inclined, through a species of intellectual myopia, to regard as the fascinating but fantastical speculations of a man whom we are now compelled, in the light of modern experience and knowledge, to admit was a prophet. But Tesla was no mere lecturer and prophet. He saw to the fulfillment of his prophesies and it has been difficult to make any but unimportant improvements in the art of radio-telegraphy without traveling part of the way at least, along a trail blazed by this pioneer who, though eminently ingenious, practical, and successful in the apparatus he devised and constructed, was so far ahead of his time that the best of us then mistook him for a dreamer. I never came anywhere near having an appreciation of what Mr. Tesla had done in this art until a very late date".

## The Wardenclyffe Tower

Starting in mid 1900, Tesla was again facing one of his perennial problems: he was running out of money for his experiments. He had been given one hundred thousand dollars by investors a few years before, but he had sunk that entire sum into his Colorado Springs laboratory. Partly as an effort to attract new investors for a new project, Tesla wrote a somewhat sensational article early that year called "The Problem of Increasing Human Energy, With Special Reference to the Harnessing of the Sun's Energy" in <u>The Century</u> magazine. Tesla's grandiloquent writing style attracted a great deal of journalistic speculation about his intentions and his projects—which was exactly what he had intended:

"Of all the endless variety of phenomena which nature presents to our senses, there is none that fills our minds with greater wonder than that inconceivably complex movement which, in its entirety, we designate as human life;

Its mysterious origin is veiled in the forever impenetrable mist of the past, its character is rendered incomprehensible by its infinite intricacy, and its destination is hidden in the unfathomable depths of the future. Whence does it come? What is it? Whither does it tend? are the great questions which the sages of all times have endeavored to answer.

"Modern science says: The sun is the past, the earth is the present, the moon is the future. From an incandescent mass we have originated, and into a frozen mass we shall turn. Merciless is the law of nature, and rapidly and irresistibly we are drawn to our doom. Lord Kelvin, in his profound meditations, allows us only a short span of life, something like six million years, after which time the suns bright light will have ceased to shine, and its life giving heat will have ebbed away, and our own earth will be a lump of ice, hurrying on through the eternal night. But do not let us despair. There will still be left upon it a glimmering spark of life, and there will be a chance to kindle a new fire on some distant star. This wonderful possibility seems, indeed, to exist, judging from Professor Dewar's beautiful experiments with liquid air, which show that germs of organic life are not destroyed by cold, no matter how intense; consequently they may be transmitted through the interstellar space. Meanwhile the cheering lights of science and art, ever increasing in intensity, illuminate our path, and marvels they disclose, and the enjoyments they offer, make us measurably forgetful of the gloomy future.

"Though we may never be able to comprehend human life, we know certainly that it is a movement, of whatever nature it be. The existence of movement unavoidably implies a body which is being moved and a force which is moving it. Hence, wherever there is life, there is a mass moved by a force. All mass possesses inertia, all force tends to persist. Owing to this universal property and condition, a body, be it at rest or in motion, tends to remain in the same state, and a force, manifesting itself anywhere and through whatever cause, produces an equivalent opposing force, and as an absolute necessity of this it follows that every movement in nature must be rhythmical. Long ago this simple truth was clearly pointed out by Herbert Spencer, who arrived at it through a somewhat different process of reasoning. It is borne out in everything we perceive—in the movement of a planet, in the surging and ebbing of the tide, in the reverberations of the air, the swinging of a pendulum, the oscillations of an electric current, and in the infinitely varied phenomena of organic life. Does not the whole of human life attest to it? Birth, growth, old age, and death of an individual, family, race, or nation, what is it all but a rhythm? All life-manifestation, then, even in its most intricate form, as exemplified in man, however involved and inscrutable, is only a movement, to which the same general laws of movement which govern throughout the physical universe must be applicable."

Tesla's publishing this article ended up attracting the attention of exactly the sort of people he needed to invest financial backing in his work. This time, it was industrialist J. Pierpont Morgan who came to Tesla's aid. Because of the extraordinary returns that George Westinghouse was still making on Tesla's induction motor patents, Tesla still seemed like a sound investment prospect to developers like Morgan. Tesla had avoided becoming entangled with Morgan in the past, fearful of the robber baron's tyrannical sway over every incorporated interest that accepted his help, but that was a long time ago; Tesla needed him now. There was a limit to Morgan's generosity, however. He was prepared to give Tesla one hundred and fifty thousand dollars in total, with a much smaller percentage of that sum as an up-front advance, and not a penny more. Furthermore, as security for this loan, he required Tesla to sign over fifty one percent of his patents. Tesla's goal might be to change the world for the better, but Morgan's goal, typically, was to make as much money as humanly possible with as little risk to his interests as he could manage. (Though, lest Morgan's investment sound less impressive than it actually was, it is worth bearing in mind that one hundred fifty thousand dollars in 1900, adjusted for inflation, is equivalent to more than four million dollars in the currency of 2016.)

Tesla's new experimental goal was worldwide broadcasting, and Morgan's financial goal was to monopolize a brand new broadcasting industry. With the money he received from Morgan, Tesla set about building a prototype industrial park. A man named James D. Warden, who owned a substantial amount of land on Long Island, gave Tesla the use of two hundred acres of it, and in gratitude, Tesla called the project community Wardenclyffe. Over the next few years, it developed into a small town, with its own post office. Its nickname, "Radio City", spoke of the inventor's hopes for what it would accomplish.

As to the broadcasting device itself, Tesla consulted with another one of his investors, famous Gilded Age architect Stanford White, as to the kind of structure it would be necessary to build in order to send signals across the world. Stanford's assistant came up with design specifications for a huge wooden tower, topped with a copper electrode one hundred feet wide. (The wooden tower itself could not be built to the necessary height, so it would sit atop the roof platform of an unremarkable but tall brick building.)

The project ran into difficulties almost immediately, the first of which being that Tesla took one look at White's specifications and knew that he could not build a tower of those dimensions with the resources he had in hand, despite the fact that the power of the signal it would undoubtedly produce would probably reach across the Pacific. Tesla began to run out of money almost as soon as he started work, and Morgan was not helpful. Morgan became even less helpful when Guglielmo Marconi made headlines in late 1900 that he had succeeded in transmitting a single letter of Morse code from a relay station in Britain to one in Newfoundland—in Morgan's eyes, this meant that Marconi had beaten Tesla, accomplishing what Tesla set out to do before Tesla managed to do it. Tesla felt otherwise; he had a more complicated goal in mind. He had sold the Wardenclyffe project to Morgan by proposing to hand him a monopoly over new radio technologies, but the real goal of the project, for Tesla, had always been to pursue his research in transmitting *power* wirelessly. As a last ditch effort to keep Morgan invested in the project, Tesla revealed this to him—but Morgan was uninterested, not least because if Tesla were successful in this, it would make obsolete some of Morgan's other business interests (which was probably why Tesla had not been up front about it in the first place.)

With the incomplete project dying on Tesla's hands, Morgan sent him a final letter indicating that he declined to take any further part in it. Tesla must certainly have been devastated; but the very night he received Morgan's letter, something quite strange happened in the Wardenclyffe park—something that Tesla would never, afterwards, fully explain. The New York newspaper *The Evening World* made this report:

> "Whatever Nikola Tesla is trying to do at Wardenclyffe, Long Island, he has succeeded in keeping the natives guessing. Some think he is trying to signal Mars; others think that he has evolved a new system of communication by electricity through the air without wires; still others believe that he has another station off in

China or Siberia, and is trying to communicate with it by electrical currents through the earth.

"Weird doings around the Tesla plant at Wardenclyffe serve to excite the inhabitants these fine nights. None of the natives is allowed to get near the bewildering stack of towers, poles and queer structures that the Tesla workmen have erected, and these same workmen are as reticent as clams. The tall electrician is seen but seldom, and when he does condescend to speak all he will admit is that his experiments have to do with wireless telegraphy.

"'Some day,' he said today, 'but not at this time, I shall make an announcement of something that I once never dreamed of.'

"For a great many years, Mr Tesla has been on the verge of making an announcement calculated to paralyze the world. In a laboratory up in Houston Street, he had a mysterious machine that poked white shafts of lightning into the atmosphere. Many men of science and finance looked at the machine and wondered.

"Similar flashes, longer and more intense, leap from the tower of the Tesla works at Wardenclyffe. The villagers sit out in front of their houses, and at intervals between batting mosquitoes from their visages speculate on the meaning of the strange lights that shoot out and appear to dissolve in the surrounding atmosphere.

"Under the tall tower there is a hole in the ground 150 feet deep. Mr Tesla admits that he shoots electric currents down this hole, and there is no doubt that he creates flashes long enough to reach the bottom of it. But there is a great deal of speculation concerning why he should want to shoot electric flashes into a hole in the ground.

"Wise-looking men of mystery who have been snooping around Wardenclyffe have been heard to say that Tesla is trying to get electricity out of the earth without the employment of artificial mediums. A man in Chicago thinks that if he can shoot a magnet into the air far enough he can accumulate electricity which can be carried to the earth on a wire. Why shouldn't Mr Tesla dig a deep hole in the ground and bring electricity to the surface? It is easier to drop a magnet into the earth than it is to fire a magnet into the atmosphere and make it stay fired!"

What sort of world changing announcement Tesla anticipated being able to make as a result of that night's bizarre, mysterious activity, the world would never know. He never explained it or made reference to it again.

# Chapter Nine: Decline and Fortune

**Financial Crisis**

Tesla was deeply in debt following the failure of the Wardenclyffe Park project. Not all of Morgan's promised money had been forthcoming, and the project had started running into debt long before it was completed. Now, with Morgan having abandoned him, Tesla had once again to scout for new investors. It was 1903, and Tesla was plagued by his creditors. Despite the fact that Leonard Curtis, who helped Tesla scout the location for his Colorado Springs laboratory, had assured him that the city power plant, of which he was a partial owner, would provide him the electricity he used in his experiments for free, the plant decided to sue Tesla for payment. The courts judged Tesla liable for a one hundred and eighty dollar fine, payable to the power company; even though the sum was a small one, comparatively speaking, Tesla did not have the money to pay it. He was eventually forced to order that his Colorado Springs laboratory be dismantled. The lumber from the building itself was sold to pay the debts, and all of Tesla's mysterious electrical equipment, which was of no real use to anyone but himself, had to be put into storage.

## Medical Innovations

One of Tesla's more or less accidental discoveries had involved the direct application of intense heat, distributed via small Tesla coils, to parts of the body that were affected by a variety of medical complaints, such as arthritis. George Scherff, Tesla's faithful assistant and manager, advised him that he was routinely receiving requests from doctors whose patients were requesting some of these devices for treatment of rheumatic complaints. Tesla told Scherff to use the laboratory to begin manufacture of some of these small medical coils, but he did not devote much personal attention to them. Sales from the coils began to generate small amounts of income.

The following year, 1906, was a bad one for Tesla in most every respect. His financial troubles continued to mount; he could not now even afford the coal necessary to run his laboratory. He was plagued by intermittent health problems, the details of which he did not describe, but it is to be assumed that they were related to the same attacks of acute sensory overload he had experienced in previous decades. George Scherff, on whom Tesla depended for nearly everything, was unable to draw sufficient pay from Tesla's business to meet his salary specifications, and began taking work from other companies that necessarily distracted him from caring for Tesla's interests. And Stanford White, the architect who had designed the Wardenclyffe Tower, was shot and killed in June of 1906 by eccentric Pittsburgh millionaire Harry Thaw, over a jealous dispute regarding Thaw's wife and White's former lover, Evelyn Nesbitt. After a famous career as an architect, the Wardenclyffe Tower would be the last structure White ever designed. Over time, lacking pay and lacking management, the workers who were necessary to running Wardenclyffe Tower disappeared to other jobs. In the end, just to pay the bill for his lodgings at the Waldorff-Astoria, Tesla was forced to sign the deed to the Wardenclyffe park over to the hotel's manager.

## The Tesla Turbine

As the decade lengthened, and the dawn of the first World War approached, Tesla's life seemed to have taken a turn for the difficult. His financial troubles met with no immediate easy solution, and on a more personal level, he seemed to feel deeply personally injured by the failures of his projects, the challenges to his patents, the attacks of the press, and the difficulties he faced in finding funding for new projects. He continued to present a dapper, sophisticated exterior, but in his personal habits, he grew increasingly eccentric. Always sensational and exhibitionistic in his communications with journalists, he began making impromptu statements that he could not justify with experiments or results; always before, he had been dropping hints about things he had already discovered, whereas now he was indulging in streams of consciousness, thinking out loud in the presence of those who inevitably printed his words, which exposed him to even more criticism and some steep humiliations. He also suffered losses of a more personal nature; his dear friend Mark Twain passed away in 1910, and his death affected Tesla deeply. And another close friend, Robert Johnson, husband of Katharine Johnson, suffered a scandal of a mysterious nature which caused him to lose his position as an editor of <u>The Century</u> magazine, which had published so many favorable articles about Tesla; the Johnsons, who were Tesla's most frequent social contacts, soon joined Tesla in fighting off debt.

But the genius's career was not yet sunk in ignominy. In 1907, Tesla, lacking funding for another laboratory, opened offices in New York, on Broadway, where he began design work on propulsion systems and vertical take off and landing for aircraft. Working off designs that he had begun in 1906, he had begun by 1910 to make serious headway on the greatest success he would enjoy in his later decades: a bladeless turbine engine that weighed under ten pounds and generated thirty horsepower. No one had ever seen or even imagined an engine that was so powerful and at the same time so small. Tesla accounted for his invention as follows:

"What I have done is to discard entirely the idea that there must be a solid wall in front of the steam and to apply in a practical way, for the first time, two properties which every physicist knows to be common to all fluids but which have not been utilized. These are adhesion and viscosity."

The Tesla turbine began to turn his fortunes around somewhat, though never to quite the degree Tesla hoped. He was fifty by this point, and unable to work at the same grueling pace as in his younger years; but the United States war department seemed to be taking a keen interest in his new invention, and Tesla's reputation benefited a good deal from the fact that he was being taken seriously by serious people once more. And a stroke of mixed good fortune came to him when J. Pierpont Morgan died later the same year. Tesla forced himself to wait for an entire month before he presented himself to J.P. Morgan, the elder Morgan's son and heir, to ask for further investments in the turbine research. Morgan was more amenable to Tesla's applications than his father had been, at first, and extended him several loans; but eventually, their relationship terminated in the same fashion that Tesla's relationship with the elder Morgan had ended, with notes on loan interest and Morgan's refusal to accept any more of Tesla's correspondence.

## World War I and the Nobel Prize

When the first World War broke out in Europe after the assassination of Archduke Franz Ferdinand by Gavrilo Princip, a member of the Serbian nationalist society, America as a whole was slow to take notice. But Tesla was himself a Serb; and Serbia was suffering intensely from the retaliation of the European powers. Tesla was approached to help head relief efforts by American-Serbian societies; but he was unable to bring himself to participate, because one of his most antagonistic scientific rivals, also a naturalized American Serb, sat at the head of the organization.

In 1915, shortly after the outbreak of the war, a curious and somewhat mysterious news report began circulating in American newspapers that Edison and Tesla had been selected as co-recipients of that year's Nobel prize in physics. When approached for comment about being selected for this honor, both Tesla and Edison professed some astonishment; neither of them had been notified of the prize committee's intention to give them the award. It was, however, not at all surprising that either man should be considered eligible, considering their contributions to science. And to give Tesla credit, while he had no difficulty thinking of various discoveries he had made that deserved a Nobel prize, he was also quick to tell reporters that Edison himself deserved the Nobel many times over.

However, in the end, the rumors came to nothing. Only a few days after the newspaper articles about Tesla and Edison's prize win started being printed, the 1915 Nobel prize in physics was awarded jointly to the father and son team William Henry Bragg and William Lawrence Bragg, of the University of Leeds, for their work in using X-rays to map the structure of crystals. The precise intentions of the Nobel committee towards Tesla and Edison has remained a mystery. Various journalists and biographers have theorized that Tesla and Edison were the original joint recipients, but that one or the other of them refused to accept the award if it meant sharing it with their obnoxious rival. If this were the case, it seems more like something that Edison would do than like something Tesla would do, if for no other reason than because Tesla is unlikely to have done anything that would jeopardize his chances at receiving twenty thousand dollars in cash. One of Edison's own biographers has theorized that Edison might have refused the Nobel simply to deprive Tesla of the much needed prize money; Edison was extremely wealthy by this point, and had no need of such a paltry sum. With the onset of late middle age and his increasingly profound deafness, Edison had grown eccentric, isolated, and emotionally cut off from his friends and family, and such a petty move would not be entirely unlike him at that point in his life.

The Nobel foundation, however, has repeatedly denied that either Tesla or Edison ever notified them of an intention to refuse a prize. They have pointed out that even if Tesla or Edison had done so, it would make no difference, because the Nobel committee does not change its decision regarding their selection of prize recipient just because it suspects the recipient will refuse to accept it. Rather, the recipient is announced, and they can then accept or refuse at their discretion. But it does seem possible that Tesla and Edison were being strongly considered by the 1915 committee, and might even have been their first choices, and that the committee's mind was changed by some factor other than a refusal—perhaps the endless newspaper articles that Tesla contributed quotes to as journals interviewed him in good faith as a Nobel winner.

## Theoreticians Versus Engineers

By 1916, science was entering a new age. During Tesla's youth, in the last two decades of the nineteenth century, the leading scientific advances were being made by inventors—that is, engineers, people like Tesla and Edison who were translating their discoveries into practical, or at least tangible, machinery. As we have discussed previously, Tesla was in many ways a theorist at heart, always deriving more satisfaction from performing experiments and chasing discoveries than from nailing down commercially viable applications for his ideas, but he was still, at heart, and engineer, someone who built things with his hands in the hopes of changing the lives of people on earth on the practical level.

In 1916, however, Albert Einstein, a German theoretical physicist, published his general theory of relativity, and seemingly overnight the entire scientific field changed from one of primarily practical, earth bound applications, to one which viewed the universe as unstable, shifting, and dynamic. Tesla was not particularly impressed by Einstein's work. He had himself devised theories about harnessing atomic power, and he was unconvinced that it was possible, or, if it proved possible, that it could be controlled. He was working on his own unified physical theory of the universe, he said. Many years later, when he was in his eighties, Tesla issued a statement outlining his dynamic theory of gravity:

"I have worked out a dynamic theory of gravity in all details and hope to give this to the world very soon. It explains the causes of this force and the motions of heavenly bodies under its influence so satisfactorily that it will put an end to idle speculations and false conceptions, as that of curved space. According to the relativists, space has a tendency to curvature owing to an inherent property or presence of celestial bodies.

"Granting a semblance of reality to this fantastic idea, it is still very self-contradictory. Every action is accompanied by an equivalent reaction and the effects of the latter are directly opposite to those of the former. Supposing that the bodies act upon the surrounding space causing curvature of the same, it appears to my simple mind that the curved spaces must react on the bodies and, producing the opposite effects, straighten out the curves.

"Since action and reaction are coexistent, it follows that the supposed curvature of space is entirely impossible -However, even if it existed it would not explain the motions of the bodies as observed. Only the existence of a field of force can account for them and its assumption dispenses with space curvature. All literature on this subject is futile and destined to oblivion."

As World War I stretched on, Tesla renewed his attempts to convince American military leaders that he had valuable scientific contributions to make to the defense of the nation. Returning to his work on robotic war boats, he urged the United States to,

"...install along both of our ocean coasts, upon proper strategic and elevated points, numerous wireless controlling plants under the command of competent officers and that to each should be assigned a number of submarine, surface, and aerial craft. From the shore station, these vessels...could be perfectly controlled...at any distance at which they remained visible through powerful telescopes...If we were properly equipped with such devises of defense it is inconceivable that any battleship or other vessel of an enemy ever could get within the zone action of these automatic craft..."


But the American military was even less interested in Tesla's ideas about robotics, and his new theories about radar, which would allow for the detection of approaching enemy craft, than they had been after his first demonstration of robotic boats during the Madison Square Garden demonstration at the time of the Spanish American War. This may have had something to do with the fact that Edison was, by now, working closely with the Department of Defense, and was as instantly dismissive of Tesla's theories about radar as he had been of his induction motor back in 1888. By the beginning of World War II two decades later, however, the usefulness of, and urgent need for a system designed according to Tesla's ideas had become apparent to American military engineers. A useful version of military adapted radar would ultimately be put into use in England just in time to enable British fighter planes to mount a defense against Nazi bombers during the Battle of Britain.

## The Edison Medal

On December 13, 1916, Tesla received one of the highest American honors in the field of science—the Edison Medal, awarded by the American Institute of Electrical Engineers. As the very name of the award must suggest, Tesla's feelings about being selected for this honor were deeply ambiguous. A fellow engineer, B.A. Behrand, had been instrumental in lobbying the award committee to bestow the medal on Tesla, but once they had agreed to do so, Behrand then had to persuade Tesla to accept it—and Tesla was not at all inclined to do so, at least at first.

Behrand was one of Tesla's most enthusiastic partisans. Outside of the inventor himself, scarcely anyone was more sensible of how much honor (and money) Tesla ought to be entitled to as a result of his scientific contributions, or felt more outrage over the fact that every leading engineer of the age had built on Tesla's inventions and theories, without reflecting proper credit and glory to the man who had originated them. But as far as Tesla was concerned, to accept a medal that bore the name of Edison, who committed so many injustices against him, was to scarcely be honored at all; he felt that every time someone received the Edison Medal, it was Edison who triumphed. Over and over again, Behrand wrote to him, urging him to accept the honor; he, at least, understood that Tesla's reputation could at this point only benefit from doing. And over and over, Tesla refused—until, finally, he changed his mind, and agreed to attend the awards ceremony.

Tesla's mixed feelings about the affair led to some strange behavior during the magnificent event which the American Institute of Electrical Engineers held in his honor. He presented himself at the banquet in all his customary resplendence of figure and dress, and behaved with good cheer and courtesy towards the Institute's members. But at one point, his nerve seemed to fail him; when the ceremony moved from the banquet hall to the auditorium, Tesla vanished. Behrand set out to look for him, thinking at first that Tesla had been taken ill and returned to his hotel. Instead, he found Tesla being watched by a crowd of spectators in the middle of Bryant Park.

Tesla had an immense affection for New York's indigent pigeon population. Most New Yorkers despised the birds as dirty disease carriers, but Tesla had been feeding and befriending them for years now. He rescued pigeons in the winter that were on the point of freezing to death; and, either because he was such a familiar figure in the park, or because of some strange animal soothing charm he possessed, the pigeons frequently took food from his fingers, and even his lips. Thus, it was to the pigeons Tesla had turned in this moment of emotional conflict just before receiving the award. When Behrand found him in Bryant Park, Tesla was covered head to toe in roosting pigeons, which fled, to his disappointment, as Behrand approached to coax Tesla back into the auditorium.

His communion with the pigeons had apparently soothed its soothing purpose. Tesla returned with Behrand, listened with composure to the highly flattering speech in which the Institute's president highlighted the main achievements of his career and made reference to the many intangible contributions Tesla had made to science and invention by inspiring other engineers. He was being honored, it was said, "for meritorious achievement in his early original work in polyphase and high frequency electric currents."

When the time came for Tesla to receive the medal, he surprised Behrand and the other attendees by making a somewhat lengthier speech than they had anticipated—a speech that gave no indication of terseness or mixed feelings, but which did credit to Tesla's graciousness and sense of courtesy, as he not only talked about his own work, but honored Edison for all that he had achieved. The following is an excerpt from his acceptance speech, and it also serves as an eloquent encomium on the philosophies that made him such an extraordinary person:

"I may say, also, that I am deeply religious at heart, although not in the orthodox meaning, and that I give myself to the constant enjoyment of believing that the greatest mysteries of our being are still to be fathomed and that, all the evidence of the senses and the teachings of exact and dry sciences to the contrary notwithstanding, death itself may not be the termination of the wonderful metamorphosis we witness. In this way I have managed to maintain an undisturbed peace of mind, to make myself proof against adversity, and to achieve contentment and happiness to a point of extracting some satisfaction even from the darker side of life, the trials and tribulations of existence. I have fame and untold wealth, more than this, and yet—how many articles have been written in which I was declared to be an impractical unsuccessful man, and how many poor, struggling writers, have called me a visionary. Such is the folly and shortsightedness of the world!

"Now that I have explained why I have preferred my work to the attainment of worldly rewards, I will touch upon a subject which will lend me to say something of greater importance and enable me to explain how I invent and develop ideas. But first I must say a few words regarding my life which was most extraordinary and wonderful in its varied impressions and incidents. In the first place, it was charmed. You have heard that one of the provisions of the Edison Medal was that the recipient should be alive. Of course the men who have received this medal have fully deserved it, in that respect, because they were alive when it was conferred upon them, but none has deserved it in anything like the measure I do, when it comes to that feature. In my youth my ignorance and lightheartedness brought me into innumerable difficulties, dangers and scrapes, from which I extricated myself as by enchantment. That occasioned my parents great concern more, perhaps, because I was the last male than because I was of their own flesh and blood. You should know that Serbians desperately cling to the preservation of the race. I was nearly drowned a dozen times. I was almost cremated three or four times and just missed being boiled alive. I was buried, abandoned and frozen. I have had narrow escapes from mad dogs, hogs and other wild animals. I have passed through dreadful diseases—have been given up by physicians three or four times in my life for good. I have met with all sorts of odd accidents—I cannot think of anything that did not happen to me, and to realize that I am here this evening, hale and hearty, young in mind and body, with all these fruitful years behind me, is little short of a miracle."

## Tesla and the White Pigeon

As Tesla entered the 1920's, he began to feel the burden of his advancing age. During his acceptance speech for the Edison Medal, he had pointed out that he came of a long-lived family, and that one of his relatives had lived to the age of 120. He himself had used to boast that he had every expectation of living until 140. But in the twenties, he began to feel differently. When the nineteenth amendment to the Constitution was passed, prohibiting the sale of alcohol, American society changed; the strictures of the first World War gave way to a carefree spirit that encouraged the flourishing of dance halls, frequented by "flappers", young women with rising hem lines, who smoked cigarettes openly, and danced with men into the wee hours of the morning. Crime flourished, as underground gangs were formed to facilitate the illegal sale of poor quality homemade alcohol in "speakeasies". Tesla, a dedicated if moderate partaker of alcohol, no longer looked forward to living into his fifteenth decade—without alcohol being available to him, it was neither possible nor desirable.

Although Tesla, almost until the end of his life, never gave up the belief that he could overcome his continuing money troubles by patenting one of his inventions for commercial use and becoming an instant millionaire, he frequently found himself penniless. He had for years been unable to afford a laboratory; he sometimes found himself unable to pay his faithful and devoted secretaries, who, rather than allowing him to use the gold in his Edison Medal to make up their back pay, instead offered to give such money as they had to *him*. Tesla's money troubles extended to his lodgings. In the last decades of his life, he had to move from one hotel to another, as he was constantly being evicted for nonpayment of his bills and forced to seek out increasingly cheaper lodgings. At one time, eviction notices distressed him because he had to find new storage facilities for his expensive experimental equipment.

Now, the insecurity of his domestic arrangements saddened him for another reason: whenever one of the pigeons Tesla fed in the park turned up ill or injured, Tesla would bring them home and nurse them back to health, or else take care of them until they died. Even when he lacked the money to pay his employees, he kept back a small sum to buy seeds for his pigeons. And when he was ill, as happened more and more frequently as he got older, the most pressing concern on his mind was to make arrangements with the hotel housekeeper to feed the convalescent pigeons roosting in cages and on the furniture in his hotel room. Unfortunately, this contributed somewhat to his having to move so often; the hotel staff were not as fond of the birds as he was, and objected to having to clean his rooms of bird droppings.

Tesla had been fond of animals throughout his life. His family in Croatia had a small farm, with geese and chickens and other common barnyard animals; and he spoke eloquently of his friendship with one of the family cats when he was a child, a cat who inspired some of his earliest thoughts about electricity. He had been petting the cat when he saw a shower of sparks—static electricity—rising off of her fur. This led to him trying to account, in a childlike way, for the existence of electricity in nature. He wrote that he had been led to wonder, "Is nature a gigantic cat? If so, who strokes its back? It could only be God, I concluded."

Of all the pigeons that Tesla nurtured in his declining years, there was one that stood out amongst them: a beautiful, snow white female, who seemed to exist in special sympathy with Tesla's moods. He wrote of his belief—possibly fanciful, possibly sincere—that he could summon the white pigeon to visit him just by thinking of her. To John O'Neill, a journalist who had befriended him, he spoke extensively of his relationship with the bird, in some of the most emotional and poignant language he ever employed:

"'But there was one pigeon, a beautiful bird, pure white with light gray tips on its wings; that one was different. It was a female. I would know that pigeon anywhere.

"'No matter where I was, that pigeon would find me; when I wanted her I had only to wish and call her and she would come flying to me. She understood me and I understood her.

"'I loved that pigeon.

"'Yes,' he replied to an unasked question. 'Yes, I loved that pigeon, I loved her as a man loves a woman, and she loved me. When she was ill, I knew, and understood; she came to my room and I stayed beside her for days. I nursed her back to health. That pigeon was the joy of my life. If she needed me, nothing else mattered. As long as I had her, there was a purpose in my life.

"'Then one night as I was lying in my bed in the dark, solving problems, as usual, she flew in through the open window and stood on my desk. I knew she wanted me; she wanted to tell me something important so I got up and went to her.

"'As I looked at her I knew she wanted to tell me—she was dying. And then, as I got her message, there came a light from her eyes—powerful beams of light.

"'Yes,' he continued, again answering an unasked question, 'it was a real light, a powerful, dazzling, blinding light, a light more intense than I had ever produced by the most powerful lamps in my laboratory.

"'When that pigeon died, something went out of my life. Up to that time I knew with a certainty that I would complete my work, no matter how ambitious my program, but when that something went out of my life I knew my life's work was finished.'"

**Tesla's "Death Ray"**

It is possible that no single subject in connection with Nikola Tesla has invited as much fascination and speculation in the minds of the general public as the rumor that he had invented a weapon which would forever change the way the war was waged on the planet, by making obsolete battles in which armies of human beings tried to kill each other close proximity with handheld weapons. An illustration in an article that appeared in The Century magazine shows an artist's depiction of the fabled weapon Tesla was supposedly working on—one that shot powerful beams of energy from fixed locations towards enemy targets. The concept of "death rays" soon entered popular culture, and became a trope in the emerging genre of science fiction, as fictional villains threatened civilization with their Tesla-inspired weapons of long-range destruction, and had to be put down by strutting heroes.

There is not much conclusive evidence for this theoretical invention of Tesla's one way or another. After a productive lull of almost six years, Tesla filed a patent in March of 1922 entitled "Improvements in Methods of and Apparatus for the Production of High Vacua".

During the Cold War, this patent was examined with great interest by both the United States and the Soviet Union, as part of their armaments program.

It is worth pointing out, however, that earlier in Tesla's life, with reference to some of the experimental results that he was producing at his Colorado Springs laboratory, he had remarked that he sometimes wondered if he had any right to build some of the things he was discovering. It seems likely that if Tesla had actually developed such a powerful weapon, this kind of ethical consideration might well have prevented him from pursuing it to the last extent. In any case, the story that Tesla developed a death ray and then destroyed it so that it could not be used to hurt anyone is now one of the most popular urban legends in circulation about him.

## Tesla, the Problem of Memory, and Death

As Tesla grew old, his powers of concentration, which had been so prodigious, and so strange all his life, began to weaken and produce distracting effects. In a letter to a poet whom Tesla befriended as an old man, he wrote of a phenomenon of memory that seemed less like a diminishing of his intellectual powers than a disintegration. He was beginning to fear that he might be suffering from the effects of a brain clot, as he was beginning to find it difficult to:

"...drive out of the mind the old images which are like corks on the water bobbing up after each submersion. But after days, weeks or months of desperate cerebration I finally succeed in filling my head chuckfull with the new subject, excluding everything, and when I reach that state I am not far from the goal. My ideas are always rational because I am an exceptionally accurate instrument of reception, in other words, a seer."

From this description, it sounds as if Tesla's old age had returned him to same problem he had overcome as a small child, when had struggled to replace the peculiar images that superimposed themselves on his vision with images of his own choosing.

In these last years of Tesla's life, he scarcely ever left his hotel room. Once relishing the finest foods and drinks, he was, in his eighties, subsisting on a diet of crackers and milk—not because he could not afford better, but because these foods best suited his palate. Though Tesla's way of living had become limited, compared to the exertions of his youth, and though he was subject to confusion and irritability, he was not senile, and he remained a creative, visionary intelligence—bending his mind to such problems as a theory of the universe that would supplant Einstein's theory of relativity. Journalists in the scientific field continued to visit him in his rooms and hang on his every word.

It was not only journalists who came to find Tesla in his retirement. As World War II loomed on the horizon, politics also came knocking. Tesla's homeland of Croatia had been folded into the new nation of Yugoslavia, and Yugoslavia was soon to be invaded and overrun by the Nazis. Serbian freedom fighters arriving in the United States sought out the most famous Serbian-American of them all, to ask for his moral support. It is believed that only a few months before he died, Tesla sent these words of encouragement back to the country where he had been born:

"Out of this war…a new world must be born, a world that would justify the sacrifices offered by humanity. This…must be a world in which there shall be no exploitation of the weak by the strong, of the good by the evil, where there will no humiliation of the poor by the violence of the rich; where the products of the intellect, science, and art will serve society for the betterment and beautification of life, and not the individuals for achieving wealth. This new world shall not be a world of the downtrodden and humiliated, but of free men and free nations, equal in dignity and respect for man."

Tesla biographer Margaret Cheney believes that Tesla began to be forgotten for a few decades after his death in part because, as a native Yugoslavian, he suffered from the hostility of western nations against those countries that became organized in the Soviet Eastern Bloc after World War II, during the Cold War. But in any case, Tesla did not live to see the Cold War, or hear of the difficulties that would be suffered by his countrymen under Soviet governments.

In the last year of his life, Tesla employed friends to carry seed out to the parks and feed the pigeons, and to bring any that were sick or injured back to his hotel room so that he could look after them. He began suffering longer periods of confusion; at one point, he believed that he had been visited by Mark Twain, who had died 25 years before, and that Twain was in financial trouble. He sent a messenger with $25 to the address of his own former laboratories on 5$^{th}$ Avenue, to relieve his friend's financial distress.

In January of 1943, Tesla passed away, seemingly in his sleep. He was 86 years old. His body was not discovered for two days, because of the "Do Not Disturb" sign he had placed on the door for the housekeepers. He had been ill for some time, but, as was his custom, had refused to see doctors. The coroner's verdict was that he had died of coronary thrombosis. The mayor of New York read a eulogy for Tesla over the radio on January 10, 1943, and his funeral was held two days later at the Cathedral of St. John the Divine, attended by over 2000 people. Tribute flooded in from European scientists; and the U.S. government paid him an even stranger tribute. After his death was discovered, the Federal Bureau of Investigation descended on his hotel room. Despite the fact that Tesla had been naturalized as a U.S. citizen many decades before, all of his paper, property, and inventions were seized as the property of a foreign alien. He had left no will. In the 1950's, Tesla's family in Yugoslavia successfully petitioned for the release of his papers, but many of them had gone missing. It is only to be conjectured what strange, powerful, and potentially world changing research Tesla had not released to the world at the time of his death, and what scientists in the employ of the American government made of them. If there was any lingering question as to the relevance of the brilliant inventor's work, the fact that the American government was apparently afraid to let it slip out of their control and into the hands of a Soviet nation probably settles it

# Works Referenced

Tesla: Man Out of Time, by Margaret Cheney (1981)


*My Inventions*, by Nikola Tesla

    http://www.teslasautobiography.com/


"Experiments with Alternate Currents of High Potential and High Frequency", by Nikola Tesla

    http://teslaresearch.jimdo.com/lectures-of-nikola-tesla/experiments-with-alternate-currents-of-high-potential-and-high-frequency-a-lecture-delivered-before-the-iee-london-february-1892/


The Inventions, Researches, and Writings of Nikola Tesla, by Thomas Commerford Martin (1894)

https://archive.org/stream/inventionsres
ear00martiala/inventionsresear00martial
a_djvu.txt

"Nikola Tesla, Dreamer: His Three Day Ship to Europe and His Scheme to Split the Earth" by Allan L. Benson (1915)

> https://play.google.com/books/reader?id
> =yvQzAQAAMAAJ&printsec=frontcover&
> output=reader&hl=en&pg=GBS.PA1763

"Nikola Tesla Promises Communication with Mars"

> https://teslauniverse.com/nikola-
> tesla/articles/nikola-tesla-promises-
> communication-mars

"Presentation of the Edison Medal to Nikola Tesla"

> http://www.tfcbooks.com/tesla/1917-05-08.htm

Made in the USA
Monee, IL
02 July 2020